Industrial Ecology

Series Editor

Syed Abdul Rehman Khan, Tsinghua University, Beijing, Beijing, China

Industrial ecology and circular economy is a peer-reviewed book series that focuses on different disciplinary approaches to waste management, sustainable practices & strategies on different scientific, societal, pyschological, technological, economic, governance, and cultural and political aspects of the ongoing and emerging debate. This primary goal of this series is to offer scientists from different school of thoughts and institutions a platform for scientific analysis and debate.

Undeniably, Industrial ecology is a rapidly growing field that systematically examines local, regional and global materials and energy uses and flows in products, processes, industrial sectors and economies. It focuses on the potential role of industry in reducing environmental burdens throughout the product life cycle from the extraction of raw materials, to the production of goods, to the use of those goods and to the management of the resulting wastes. Industrial ecology is ecological in that it (1) places human activity—industry in the very broadest sense—in the larger context of the biophysical environment from which we obtain resources and into which we place our wastes, and (2) looks to the natural world for models of highly efficient use of resources, energy and byproducts. By selectively applying these models, the environmental performance of industry can be improved. Industrial ecology sees corporate entities as key players in the protection of the environment, particularly where technological innovation is an avenue for environmental improvement. As repositories of technological expertise in our society, corporations provide crucial leverage in attacking environmental problems through product and process design.

More information about this series at https://link.springer.com/bookseries/16605

Fatima Ezahra Touriki · Amine Belhadi ·
Sachin Kamble · Imane Benkhati

Sustainable Excellence in Small and Medium Sized Enterprises

Continuous Improvement Approaches that Matter

Springer

Fatima Ezahra Touriki
Industrial Management
Cadi Ayyad University
National School of Applied Sciences (ENSA) Safi
Safi, Morocco

Sachin Kamble
Strategy (Operations and Supply Chain Management)
EDHEC Business School
Roubaix, France

Amine Belhadi
Cadi Ayyad University
Safi, Morocco

Imane Benkhati
Industrial Management, ENSA-Safi
Cadi Ayyad University
Safi, Morocco

ISSN 2730-5775 ISSN 2730-5783 (electronic)
Industrial Ecology
ISBN 978-981-19-0370-0 ISBN 978-981-19-0371-7 (eBook)
https://doi.org/10.1007/978-981-19-0371-7

This Springer imprint is published by the registered company Springer Nature Singapore Pte Ltd.
The registered company address is: 152 Beach Road, #21-01/04 Gateway East, Singapore 189721, Singapore

Introduction

Recent studies have clearly highlighted the significant role of small and medium-sized enterprises (SMEs) as the driving force behind many of the world's leading economies (Belhadi, et al., 2018a). Moreover, according to statistics from the Organization for Economic Co-operation and Development (OECD), SMEs represent more than 99% of all businesses in the world (Marchese, et al., 2019).

However, despite their essential role in the economic growth, SMEs are not very innovative in terms of managerial practices, unlike their large counterparts. This category of companies is the most subject to the impacts of dynamic changes in their environment. For example, the global health crisis linked to the coronavirus pandemic, has upset the usual operating methods of all SMEs in different sectors. The impacts touched mainly the economic and social aspects of sustainability. Economically, the world has witnessed business bankruptcies, reductions in activities, various bottlenecks in supply chains, and growing debts. The social level was aggressively impacted mainly because of the unemployment growing rates and confinements procedures.

Moreover, companies found themselves, unprecedently, facing a new challenge/opportunity keeping up with the new era of smart technologies namely industry 4.0. Being part of the journey of digitalization is no longer a choice neither a luxury, it has become a necessity in order to remain competitive, improve efficiency, agility, and productivity.

Indeed, today, digital technologies offer several opportunies:

- Time saving for reporting thanks to the IT tools integrated into the management of the company's various processes (maintenance, production, quality, etc...);
- Simulation possibilities;
- Data analysis for decision-making;
- Organizational learning...

These challenges can be seen as mega forces that require the introduction or the strengthening of the company's management methods to stabilize and improve performance. Companies are thus called upon to readjust their strategies and make a vital decision that is often difficult: to maintain the usual mode of governance and operation or to opt for an efficient transformation with a wider spectrum of objectives (economic, social, and environmental), while benefiting from various tools, in particular those offered by industry 4.0. This strategic decision will lead SMEs either to decline if the current mode of operation is no longer adapted to the context of the organization, or to prosperity and a new state of stability.

Faced with this situation, the choice of the sustainable excellence approach has emerged as a relevant response to the aforementioned challenges. Despite being highly popular in the scale of large enterprises, "Managing sustainably" is an emerging approach for small structures. The convergence between economy and ecology is now widely accepted in the context of globalization. As such, SMEs are called to explore their operational and external performance while taking into consideration stakeholder expectations and environmental matters and social issues. Being excellent sustainably does not mean opting for a strategy of imitation of successful companies, but rather an adaptation strategy based on a thorough knowledge of both internal resource capabilities and external opportunities and risks.

In practice, sustainable excellence can be seen as a fusion between operational excellence programmes promoted by several approaches such as TQM, lean, Lean six Sigma and the concept of sustainable development through environmental management, circular economy and corporate social responsibility. It is a dynamic and interactive process that enables the company to achieve high long-term performance while harmoniously integrating the demands and expectations of stakeholders expressed in economic, social, and environmental terms.

Building on the above, this book aims to gather current experience and knowledge on the implementation of sustainable business excellence in the context of SMEs. The book uses an empirical and practical approach to address these objectives, which are underdeveloped in the literature. It could therefore be a relevant reference for SME managers seeking to manage their operations in a sustainable, efficient, and resilient manner.

References

Belhadi A, Touriki FE, ElFezazi S (2018a) Lean implementation in small and medium-sized enterprises in less developed countries: Some empirical evidences from North Africa. J Small Bus Manage 56(sup1):132–153.

Marchese M, Giuliani E, Salazar-Elena JC, Stone I (2019) Enhancing SME productivity: Policy highlights on the role of managerial skills, workforce skills and business linkages. OECD SME and Entrepreneurship Papers, Issue 16.

Contents

Chapter 1
Sustainable Excellence and Continuous Improvement Approaches in SMEs

Objectives

1. Define sustainable excellence and its importance in improving performance,
2. Explain the continuous improvement approaches: Models of excellence, Circular Economy, Environmental Management, Lean Manufacturing and Social Responsibility.
3. Explain the implementation approaches.

1.1 Introduction

In the current context of globalization, sustainable excellence is a logical path for any company that aware of its social responsibility and its impact on the environment. Indeed, this awareness of the current production and consumption patterns having a strong impact on the planet's ecosystems, leads SMEs to face moral, legal and/or economic obligations to preserve their activities while maintaining a viable production system in the long term.

This chapter presents the continuous improvement approaches, from the most traditional to the most innovative, in terms of sustainable excellence, allowing SMEs to control and optimize their performance. In other words, it will address the different options available to companies.

Given the evolution of the economic, regulatory, technological, social and ecological context and the adaptation efforts that companies must rapidly deploy, it is also important to know the steps to follow for the adoption of such improving approaches. Indeed, once the strategic orientations are decided, the company leaders and their collaborators draw up the path to follow in order to reach their objectives. For this reason, some deployment methods will also be presented. In this context, a presentation of what is sustainable excellence and what are the approaches covered by this concept are de main objectives of the following sections.

F. E. Touriki et al., *Sustainable Excellence in Small and Medium Sized Enterprises*, Industrial Ecology, https://doi.org/10.1007/978-981-19-0371-7_1

1.2 Sustainable Excellence

1.2.1 History

From a purely linguistic point of view, the expression "excellence" refers to the notions of superiority and perfection, which can be translated at the company level by the existence of superior performance compared to its peers. In this sense, Nha NGUYEN (2006), defines excellence as: "Maintaining mobilization and performance in order to become and remain the best, to surpass oneself, to leave a lot of room for innovation, to deploy oneself, to manage in a network and to share knowledge".

The term "excellence" here refers to a dynamic process whose goal is to continuously improve performance in order to maintain this position of predominance. This is achieved through the availability of resources, skills and innovation capabilities. Without forgetting that the excellent company necessarily follows the evolution of its context, it is called to be in synchronous mode with the recent digital technologies, to exploit them for a better management of the competences and the resources including the knowledge. Therefore, the concept of excellence represents both an end and the possession of means to achieve it.

It should also be noted that through this definition of excellence, collective intelligence is at the center of actions. It goes without saying that the quest for excellence necessarily involves the construction of skills that the company must mobilize and strengthen in order to achieve sustainable performance over time. According to Amit and Schoemaker (1993), this construction of skills, whether on an individual or collective scale, is based on the accumulation of knowledge and collective learning.

These two characteristics (cumulative and interactive) also apply to the notion of excellence. ROUBY AND SOLLE (2001) emphasize that the dynamic interaction factor of organizational competencies at different levels if the company wishes to achieve excellence and carry out its business over time while meeting the expectations of its partners.

Others, some authors such as BERGERY AND DEJOUX (2006) consider excellence as "the result of a ranking and not as an intrinsic characteristic". The assessment of excellence implies the comparison of the organization's mode of operation with existing best practices. This idea, which is not completely false, reflects the evolution of the notion of excellence over the last four decades:

In 1982: the publication of the book "In Search of Excellence" by Peters and Waterman (1982) resulted in a list of the most successful companies of McKinsey's clients. The study adopted a method developed by McKinsey based on the 7S: Strategy, Structure, Systems, Style, Staff, Skills, Shared Value. Their work allowed them to identify, among other things, the eight character traits of successful organizations, which are: (1) Value action first and foremost, (2) Listen to the customer, (3) Encourage autonomy and innovation, (4) Base their productivity on the motivation of their teams, (5) Rely on a key value, (6) Stick to what they know how to do, (7)

Implement a simple and light organization, (8) Combine flexibility and rigor. The study also shows that excellence is a set of practices that affect several levels of the company: operational, cultural, strategic and managerial.

In 1992: the European Foundation for Quality Management published the EFQM excellence Model, based on the idea that access to excellence requires comparison with a reference. It is both a guide to an effective management system and a tool for internal diagnosis. Indeed, it proposes an architecture for excellence based on nine criteria, five of which constitute the "Enablers": Leadership, People, Strategy, Partnership & Resources, Processes, Products & Services. These enablers must be mastered and optimized to achieve successful results in terms of: People, Customer, Society and Key Results (Gomez-Lopez et al. 2016). The model encourages companies to become more agile and responsive to their context, and thus sets out eight fundamental concepts of excellence, namely: (1) Leading with Vision, Inspiration & Integrity; (2) Adding Value for Customers, (3) Creating a Sustainable Future, (4) Developing Organizational Capability, (5) Managing with Agility, (6) Harnessing Creativity & Innovation, (7) Succeeding through the talent of people, (8) Sustaining Outstanding results. (Jabnoun 2019).

In 1994: once again, in the logic of external evaluation, the publication of standards Iso 9001, Iso 9002 and Iso 9003 set the roadmap for companies wishing to embark on the path of Quality Assurance and by extension to excellence through several requirements mainly organizational. Note that ISO 9004 is not a standard in the sense that it is not certifiable, it is not a model of excellence, but rather a document providing recommendations for a better application and compliance with the requirements of ISO 9001. In its appendices, it proposes a system of self-assessment of the effectiveness and efficiency of the organization's practices, thus providing an assessment of the level of maturity of a quality management system (see Table 1.1). The objective is to provide the organization with evidence-based guidance on where to invest resources for improvement (ISO 2000).

At that time, ISO 9004 offered managers a valuable tool for evaluating performance and structuring the organization's action plans.

From the year 2000: the evolution of the ISO 9001 standards, pushed companies in their strategic planning to take into consideration in addition to the customers' requirements, those relevant to the "Interested or Stakeholders". This includes suppliers, subcontractors, shareholders, employees or simply society in general through its requirements in terms of respect for the environment. This evolution has led organizations to comply with several ISO standards in different areas: Environment (ISO 14001), Occupational Health and Safety (ISO 45001), Social Responsibility (ISO 26000).

Table 1.1 Maturity and performance levels (Extract from the international standard ISO 9004: 2000 quality management systems-guidelines for performance improvement

Maturity level	Performance level	Guidelines
1	Informal approach	No obvious methodical approach, no, poor or unpredictable results
2	Reactive approach	Methodical approach based on problems or corrections; minimal data available on results regarding improvement
3	Formal stable system approach	Methodical process-based approach, early stage of systematic improvements, available data on compliance with objectives and existence of improvement trends
4	Enhanced continuous improvement	Improvement process used, good results and strong improvement trends
5	Optimal Performance	Deeply rooted improvement process, proven optimal competitive benchmarking results

1.2.2 Sustainable Excellence

As a result of this evolution of standards and references, the concept of excellence is now complete and more complex. It seeks to satisfy the relevant needs and expectations of the various stakeholders and progressively incorporates the dimensions of sustainable development. This extension of the concept of excellence towards sustainable development has given rise to "Sustainable Excellence". The meaning given to the term "sustainable" in this expression is not necessarily linked to time. In fact, it goes beyond the simple acquisition of sustainable performance to achieve a long-term competitive advantage. Our representation of the notion "Sustainable" refers to the concept of "Sustainable Development" in its broadest acceptance. The idea is not to question the economic purpose of the company, but rather to associate a certain number of decisions and complementary actions that place the company in a register of economic, environmental and social development.

Spread on a large scale since 1987 by the report of the Brundtland Commission, Sustainable Development refers to "*development that meets the needs of the present without compromising the ability of future generations to meet their own needs*". It emerges that this concept refers to three parallel dimensions according to (Elkington 1997):

- **The social dimension:** constitutes the objective of the concept in the sense that current development must respect the needs of future generations but also those of the present in terms of employment and social equity.
- **The economic dimension**: represents the engine and the means to be used for sustainable development. (Growth and profitability)

- **The environmental dimension**: indicates the condition of sustainable development, based on the postulate that industrial activities have direct and indirect effects on natural resources and the functioning of ecosystems.

From this perspective, "sustainable excellence" encompasses not only the vertical logic of climbing the ladder to become the best, but also a horizontal logic that allows the development of mutually beneficial partnerships with stakeholders and/or new opportunities for collaboration that promise to create value.

It should be noted that there is no instantaneous relationship between the three dimensions of sustainable development. The economic dimension follows a short-term profit logic, while the environmental and social dimensions follow a long-term logic. Moreover, the question of links and correlation between these three dimensions is still a subject of debate in the scientific and practical communities. The literature review conducted by Allouche and Laroche (2005), shows that most studies (71%) show a positive link between financial performance and social performance, regardless of the nature of the measures of social performance and financial performance. The level of financial performance at least partially determines the level of social responsibility.

This relationship is strengthened and consolidated by the presence of more and more sustainable development regulations in several countries, notably between the environmental and economic dimensions. For example, the application of the Polluter Pays Principle accentuates the link between financial and ecological performance. For example, in France, the adoption of the NRE law (New Economic Regulations) from 2002, obliges companies to report in their annual report, the way they integrate the social and environmental consequences of their activities (André et al. 2011). This obligation of information must deal with four components according to (Mauléon and Silva 2009):

1. The objectives and measures taken in terms of environment, employment, social policy and social protection,
2. Social information: changes in the workforce, work organization, remuneration, health and safety conditions, etc.
3. Information on the company's social impact: contribution to the socio-economic development of the region, relations with environmental protection associations, local authorities, etc.
4. Environmental information: consumption of water, raw materials and energy, measures taken to improve energy efficiency, etc.

1.2.3 The Core Characteristics of Sustainable Excellence

Sustainable excellence mainly reflects a state of mind in perpetual awakening propelling the company towards new ways of improvement and/or mastery of existing practices. It conveys a beneficial and positive transformation based on the adoption of fundamental values and real convictions touching several aspects:

- **Strategic planning:** Achieving excellence and becoming the leader in one's field are challenging objectives for many companies, regardless of their size. Small and medium-sized businesses in particular feel the need to evolve and sustain their competitive position. To do so, strategic planning is the most essential ingredient in this recipe for success. Numerous studies have demonstrated a positive link between strategic planning and performance (Cho and Pucik 2005; Carmeli 2004; Ketchen et al. 2004). Consistently, companies that plan their strategy have higher financial performance than those that do not plan their strategy. Strategic planning is the formal identification of objectives and the choice of approach to achieve them. Formalization is essential, because it allows communication with the collaborators in the continuous improvement project, and on the other hand, it avoids dispersion thus allowing to focus on the essential (Robbins et al. 2014). To facilitate this planning, we propose a description of possible approaches and methods to achieve sustainable excellence. The continuous improvement approaches explained will address the environmental, operational, social and organizational aspects.
- **Stakeholders:** the notion of stakeholder is omnipresent in sustainable excellence. If a company wishes to achieve long-term performance, it is fundamental to work on its network of partnerships and collaboration, whether internal or external. Internally, the acquisition, mobilization and maintenance of a high level of competence is an absolute necessity. Taking the human aspect into account in the management process towards excellence will be reflected in the results in terms of social responsibility, health and safety at work and in the operational system: competent and committed internal employees will have a positive impact on operational and economic performance. Moreover, as Dervitsiotis (2004) points out, "So, rather than looking at the human work community of practice as costs or 'resources' to be used, the knowledge era demands that they must be dealt with as the only means of developing sustainable organizations". Externally, the company must be aware that it does not operate in an isolated environment, on the contrary, it is part of an extensive and complex ecosystem. The organisms that make up this ecosystem interact with each other and can promote or limit the proper functioning or development of the company. The company must therefore locate and identify the actors potentially in close interactivity with these activities and then involve them in such a way as to establish a mutually beneficial symbiotic relationship. This will allow the company to better understand the scope and extent of its environment, and its position in its sphere of influence. Inserted in the circular economy approach allows for benefits in terms of value creation.
- **Permanent organizational learning**: according to Koenig (1994), organizational learning is defined as "a collective phenomenon of acquisition and development of competencies that modifies situations in a profound and lasting way". This practice, essential in organizations, is very much in demand, especially in the current context characterized by changes in the economic, regulatory, technological and ecological environment. Moreover, the survival of the company depends on its ability to respond faster than others (Forman et al. 2007). The best example is the "Toyota" company, also known as the "learning company" Riztman et al.

(2013) in which organizational learning is a shared motto among the different employees. To improve process productivity and reduce waste, Shiego Shingo, one of its engineers challenged his employees to "innovate and find solutions to speed up the production rate while eliminating unnecessary activities". This experience showed that organizational learning remains the best way to increase the performance of the operational team in terms of skills but also in terms of flexibility and reactivity to changes in its context. The organizational dynamic will be in permanent learning mode. This concerns all levels of the company, all personnel, takes all possible forms and develops in interaction with the outside world (Forman et al. 2007). The means to promote it are the integration of ICT tools (information and communication technologies) but also via learning experiences in the field (in the form of Kaizen projects or others). This last mode is very effective and beneficial both for improving performance and for reinforcing the feeling of achievement and success among employees. With the accumulation of experiences, the sharing of information, knowledge and know-how, the organization will develop collective and competitive intelligence.

- **Operational excellence of the processes:** the management by process approach conveyed by the Iso standards allows at the same time to have a vision on the various activities of the company but also an effective and efficient management of these activities in a global and systemic framework. This approach is fundamentally interesting insofar as any change project will be facilitated by a coordinated improvement between the different processes of the system. The means to achieve this excellence remain Total Quality Management, Lean & Lean Six Sigma, methods applied to environmental management (Life Cycle Assessment; Eco-design…).
- **Information System and Technology of Industry 4.0**: It should be recalled here that the evolution of information systems and technological change are strongly present in the current context of companies. This particularly requires attention and willingness on the part of managers but also an effort of appropriation and integration on the part of employees. Especially since Industry 4.0 technologies offer very attractive benefits in terms of operational control of processes, collaboration methods and relations with stakeholders. The development of an internal information system is strongly recommended to facilitate organizational and collective learning, and to monitor and control operational performance (Fig. 1.1).

1.3 Models of Excellence

Very often, excellence models, in the literature, are strongly associated with the TQM concept "Total Quality Management". Indeed, for a long time, companies have been opting for approaches based on quality as a strategic differentiation factor in a highly competitive environment. Moreover, the benefits of the (TQM) approach impact both the internal activities of the company as a means of optimization and improvement, and the external activities through the recognition of the performances obtained.

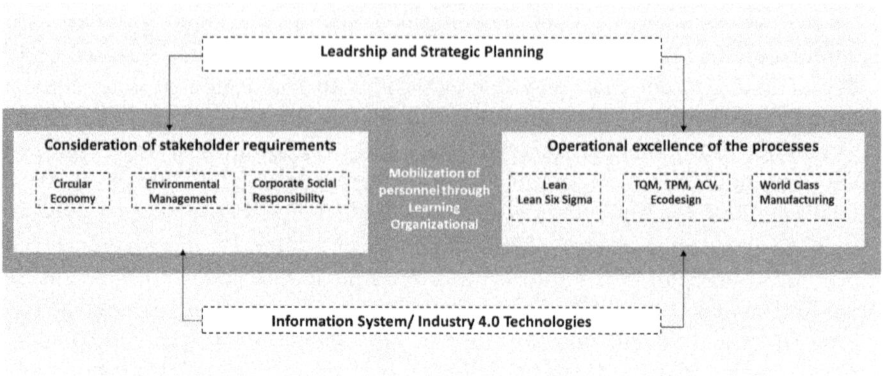

Fig. 1.1 The fundamental characteristics of "sustainable excellence"

In this sense, in order to evaluate companies according to the development of their TQM system principles and to reward best practices, several models were created. The first "Deming Award" was created in 1951 in Japan, then several countries have established this practice of excellence awards such as the Malcolm Baldrige National Quality Award (MBNQA) model in the United States, the European Quality Award (EFQM excellence model) and even newly industrialized countries, such as the Republic of Singapore, created the Singapore Quality Award (SQA) in 1993 (NGUYEN 2006). This spread of excellence awards around the world demonstrates that quality awards have become an important reference tool for public and private sector organizations (Löffler 2001).

Although the TQM approach is very popularized by quality awards, other models of excellence have been identified and classified according to (Sharma and Kodali 2008) into three broad categories: award-based models, academic/researcher-based models and consultant-based models.

In the following, models of managerial, operational and sustainable excellence with a clear stakeholder orientation, a focus on the role of leadership and the human aspect in the success of the approach will be presented. These models, often referred to as holistic approaches, provide the opportunity to discover appropriate answers to the daily concerns of companies.

1.3.1 The European EFQM Model

Being a holistic approach, the application of the EFQM model (European Foundation for Quality Management) establishes coherence and alignment between the company's objectives and its operational and managerial practices. It is part of a global and integrated vision affecting all activities and processes. First, its application allows an analysis and evaluation of the current state of the company through the identification of points to exploit (strengths) and points to improve (weaknesses). Secondly, it guides the organization in defining objectives, developing a global strategy to achieve them and creating a set of action plans to integrate and coordinate activities. Since it is a model of excellence, and as explained earlier, it is as much about the ends (what to do) as it is about the means (the resources to achieve the results). Finally, the model encourages a culture of measurement, through the recording of results and performance. According to (Dadfar et al. 2015) found that performance based on the EFQM model correlated positively with benefits derived from increased exports.

The structure of the EFQM model is based on three axes and that reflects the construction logic of the excellence approach in practice. First, the "Management" axis focuses on the company's raison d'être, answering the question why does this organization exist? Why does it choose this strategy? In the "Execution" axis, the company is led to reflect on how to proceed in order to achieve its objectives, in other words, how does it plan to deploy its strategy? Finally, the last axis "Results" questions the company on its current and expected performance (EFQM 2021). This logic is reflected in Table 1.2, which summarizes the axes of the model, their purpose and their associated criteria.

It is a world-renowned practical framework that can lead to external recognition if the organization achieves sustainable and outstanding performance. The assessment follows a 1000-point scoring system with 200 points for management, 400 points for execution and 400 points for results.

1.3.2 Management System Standards

The objectives of sustainable excellence are not fundamentally different from those of ISO standards that deal with management systems or even models of excellence such as the EFQM (Forman et al. 2007). The latter explain that for each of these standards, the aim is to facilitate the implementation of a management system that strengthens:

- The sustainability and competitiveness of the company,
- The control of global risks, especially environmental,
- The satisfaction of interested parties.

Consequently, standards are above all tools and methodological guides available to their users (public or private) to harmonize practices and improve their interoperability (Lefebvre 2018). System standards such as ISO 9001/14001 and 45,001, which deal respectively with quality, the environment, and occupational health and safety, enable organizations to pursue best practices in order to meet the threefold objective of sustainable excellence, especially when they are approached in an integrated manner.

The advantages of improvement approach according to Iso standards are many and varied, in addition to the attractiveness and economic advantage:

Table 1.2 Guidelines and criteria of the EFQM model (EFQM 2021)

Focus	Goal	Criteria
Direction	For an organization to achieve and sustain outstanding results that meet or exceed the expectations of its stakeholders it: – Defines an inspiring Purpose, – Creates a Vision that is aspirational, – Develops a Strategy that is centered on Creating Sustainable Value, – Builds a winning culture This Direction setting prepares the way forward for the organisation to be seen as a leader in its ecosystem and well positioned to execute its plans	1. Purpose, vision strategy 2. Organizational Culture & Leadership
Execution	The Direction setting as outlined above, prepares the way forward for the organization, but it then needs to execute its Strategy effectively and efficiently, ensuring that it: – Knows who the stakeholders are in its ecosystem and engages fully with those that are Key to its success, – Creates Sustainable Value, – Drives the levels of performance necessary for success today and, at the same time, drives the necessary improvement and transformation if it is to be successful in the future	3. Engaging Stakeholders 4. Creating Sustainable Value 5. Driving Performance & Transformation
Results	What the organization has achieved in relation to what has been described in the Direction & Execution sections, including the forecast for the future. In practice we find that an outstanding organization provides results data for: Stakeholder Perceptions Creating Sustainable Value Driving Performance & Transformation	6. Stakeholder Perceptions 7. Strategic & Operational performance

First, as the Iso standards stipulate, understanding the organization's context is an essential prerequisite for propelling the company towards an upward trajectory. (Villain 1990) emphasizes the need for a company to gather information about its environment and to manage these incoming and outgoing information flows in an era where competition is increasingly crucial. Monitoring the environment consists in detecting trends and prospects in order to make the most of them for future action.

According to these standards, a company's strategic thinking starts with an internal and external analysis of the factors influencing its activities and processes. The external analysis calls for strategic intelligence activities, which is defined as: "*the informational process by which the company detects and processes the signals announcing events likely to influence its sustainability. The goal... is to allow the company to reduce its uncertainty, in particular by anticipating the ruptures that can occur in its socio-economic and technological environment*" (Lesca and Caron 1995). When it is well conducted and practiced, it will allow the company to access competitive intelligence[1]. The author, Henri (DOU 1995), considers monitoring to be a practice that allows one to have the intelligence of one's environment through the implementation of one's own observation and reflection capacities and through the organization of a relevant system of intelligence creation. The goal is to monitor, understand the environment and analyze its implications in order to obtain information of great added value, which allows, for example, to know the strategic intentions of a competitor from the weak signals collected.

Indeed, the knowledge of the external environment conveys a competitive advantage when the observation of the external world is done in a panoramic and selective way. It must be sufficiently broad to situate the environment in its globality: technological, regulatory, economic, competitive, societal, ecological and thus appreciate the nature and scope of the major phenomena that emerge. It must also be selective because there are critical factors to be monitored more closely than others because they can have a major impact on the company's important decisions and strategic orientations (regulations for example). The time factor is also crucial, as Morin (1985) points out, companies are currently challenged on the ability of their management to detect threats in a timely manner, transforming them, if possible, into innovative opportunities for their own development.

The establishment of a reasoning and a methodology for monitoring the environment will allow to detect risks to avoid them and opportunities to seize them. Without this tool, the company will remain isolated from its environment and will not be aware of the changes that occur. This situation can cause the ruin and bankruptcy of some. The example of Kodak is very striking in this context, despite its position as world leader at the time, it was destabilized by the technological evolution of its market. It did not seize the opportunity to engage in the manufacture of digital cameras, unlike other competitors known in the field of consumer appliances (Sony, Phillips) invested in this very promising market. As a result, after posting a loss in the first half of the year representing 3.2% of its turnover, Kodak announced a new redundancy plan providing for 25,000 job cuts instead of the 15,000 planned in 2004 (Fernez-Walch and Romon 2006).

In addition to the external analysis, it is necessary for the company to highlight these capabilities in terms of internal resources and skills, which are often the basis of any sustainable competitive advantage. This will allow the company to establish an adequacy and a correspondence between these capacities and resources and its ambitions.

Then, the Iso standards insist on considering the expectations and requirements of the stakeholders, all categories included, which allows the company to be part of this sustainability logic.

Also, the PDCA logic installs within the company a simple but effective operating mode based on a reflection before the action by planning and an evaluation after the action by the audit and the control.

Another advantage lies in the fact that, by nature, standards explain the requirements to be met without specifying the how. This implies research and effort on the part of those who implement them, thus promoting organizational and collective learning as well as creativity and innovation.

Finally, another advantage is the role of facilitating compliance with regulations. During the implementation of standards (9001, 14,001 and 45,001), it is an opportunity for the company to comply with the laws in force (labor code, laws on waste, pollution …). By means of a monitoring tool, this time regulatory, allowing him to identify the applicable laws and verify their application and applicability on the ground.

These comments are supported by an AFNOR Certification study (carried out in 2008 on QSE management systems) in (Lefebvre 2018) which lists and ranks the benefits of these systems according to the responses obtained:

- Improve the overall efficiency of the organization,
- Improve the company's image,
- Meet and exceed regulatory requirements,
- Satisfy a customer obligation and/or meet a Group requirement
- Prevent or remedy malfunctions.

Indeed, opting for an integrated management system (IMS) is interesting on several levels. Firstly, the three standards present the same concepts and principles of management, the same structure (governed by HLS: High Level Structure) which favors a global coherence (See Fig. 1.2). Secondly, the company wins in terms of time and cost when it invests in a joint application of the standards. Finally, considering the expectations and requirements of the various stakeholders is the strength of the SMI.

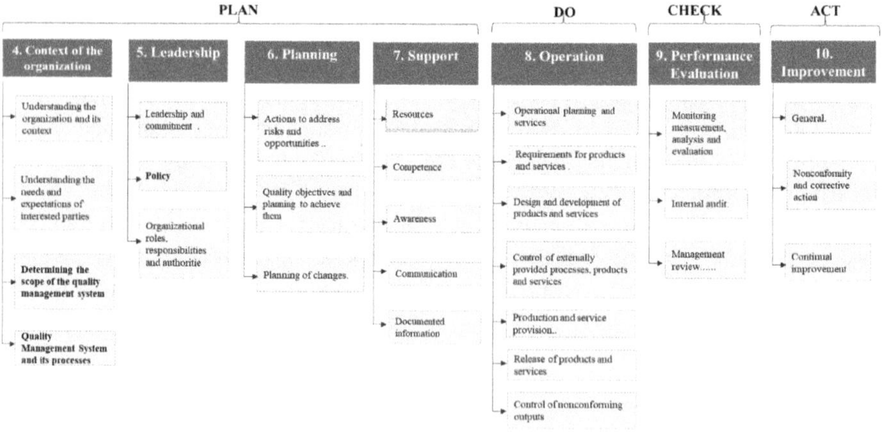

Fig. 1.2 The requirements of the Iso 9001 Version 2015 standard according to the PDCA logic

Companies that have opted for excellence models based on Iso standards have mainly external motivations such as responding to customers' wishes or for external representation purposes Heras-Saizarbitoria et al. (2006).

1.3.3 World Class Manufacturing

Unlike the EFQM model which requires external recognition, World Class Manufacturing (WCM) are development models built and formalized internally by the companies themselves (especially multinationals or very large companies). As it is not a label or an award, companies adopting it are expected to perform better than their counterparts in the market, which may be competitors or subsidiaries of the same multinational.

The term "World Class" was introduced in 1986 by Richard J. Schonberger's book published by The Free Press. Since then, several models have emerged, with a strong presence of the automotive industry, such as CHRYSLER, FAURECIA…but the pioneer in the field remains the Toyota Production System (TPS). Also known as "Lean Manufacturing".

Despite their multitude, the models nevertheless have some common characteristics:

1. The objectives: Zero defects, Zero breakdowns, Zero accidents, Zero. Although this may seem utopian, the idea is to put in place a set of locks that will allow us to "Tend towards" the four Zeros,
2. Global approaches integrating all the activities and processes of the company,
3. The staff is the driving force of the model, it is regularly trained and called to carry out actions in its field of work.
4. The principle of evaluation is strongly present in a systematic way via audits or grids built according to the requirements of the model. This allows the work teams to be involved in a dynamic process through the action plans resulting from these evaluations.

In practice, WCM consists of the combination and association of several concepts and approaches (see Figure…) with the aim of the improvement and the harmonious and sustainable development of the company. The five key approaches that compose it are (Hennion and Makhlouf 2016):

- Just in time (JIT) to synchronize production with the customer,
- Total Productive Maintenance (TPM) to ensure equipment reliability,
- Total Quality Control (TQC) to ensure product quality,
- Total Industry Engineering (TIE) for productivity,
- Total Quality Management (TQM), for problem solving and continuous improvement.

Japanese Professor Hajime Yamashima (Yamashina 1995) has contributed to the understanding of this integrated production system by explaining its components.

Presented in the form of an ancient temple, supported by pillars constituting the technical component which are based on managerial foundations. This multitude of criteria demonstrates the willingness of companies committing to WCM type models to provide answers on several levels both internal and external. Each of the pillars is implemented and deployed according to a seven-step process (Hennion and Makhlouf 2016):

- **Reaction phase:** consisting of three steps (1. Identify the problem to be addressed, 2. Detect where it appears, 3. Prioritize the problems according to the costs of corrective actions).
- **Prevention phase**: consisting of two steps: (4. Analyze the solutions and estimate the costs, 5. Choose the best method to avoid the recurrence of known problems).
- **Proaction phase:** established in two steps (6. Implement the solutions with rigor and evaluate the results, then compare them with the initial objectives, 7. Based on a risk analysis, implement preventive actions to avoid the appearance of new problems).

It is clear from the pillars that make up WCM that it is based on the use of tools linked to Total Production Maintenance or cost and loss analysis, on approaches that take into consideration the expectations of stakeholders, such as the environment and safety pillar, and on organizational values such as employee involvement. This heterogeneous mix of techniques and methods makes WCM a global and federative approach to continuous improvement. Global in the sense that it allows the reconfiguration of the company's various processes through best professional practices (logistics, quality, maintenance, production, security, etc.) with a view to reducing costs. It is also unifying, since it is a culture applied at all levels of the organization and by all teams. A culture that is disseminated and practiced essentially through organizational learning via activities that focus on the development of individual and collective employee skills. Moreover, only learning organizations can survive and succeed in a competitive market.

The model also advocates respect for stakeholders through its pillars related to safety (for employees), the environment (for society), quality control (for the customer) and the development of employee skills.

It is a complete and complex model that requires rigor and perseverance. Although it offers a multitude of approaches, tools and values, it suffers, as do the other models mentioned above, from the lack of an information system.

1.4 The Mode of Implementation of Sustainable Excellence

Being sustainably excellent does not mean opting for a strategy of imitating successful companies. Rather, it is a strategy of adaptation based on a thorough understanding of both internal resource capabilities and external opportunities and risks.

From the models of excellence detailed above, one major difficulty that SMEs may encounter is the lack of an implementation guide or the lack of uniformity in the implementation procedure. It is true that it is difficult for SMEs to find easily accessible repositories that provide information on the sequence to follow and the chronology of the steps to implement. Several questions come to mind: should one engage in the simultaneous or sequential implementation of the main concepts cited in a model? should one start with an initial training activity or move forward without training relying on learning by doing? Are there tools to facilitate the appropriation and integration of the excellence models? How to manage the resistance to change induced by these new projects? These are all questions that deserve clarification for SMEs managers for a better appropriation of Sustainable Excellence.

Therefore, the explanation of a deployment process is of crucial importance to facilitate the adoption of excellence models, this is especially true for SMEs that suffer from a lack of resources and cannot afford the support of consultants. This is especially true for SMEs that lack resources and may not be able to afford the support of consultants. Once the decision has been made, this is not a one-time action, but a process that takes place over time... These models give an idea of the implementation methodology through at least three main steps (in a very simplistic and general way). Firstly, the analysis of the existing situation via a diagnosis allowing the identification of opportunities, then the construction of action plans specifying objectives and methods to be followed in order to achieve them, and finally the monitoring of the performance obtained. As far as the first step is concerned, these models all have one thing in common: they offer their users a self-assessment tool based on several criteria, which will serve to identify the sources of vulnerability and consequently the areas for improvement for the company. On the other hand, taking the example of Iso standards, they often simply indicate what needs to be done to achieve the quality improvement objective, but do not explain how to do it (Pfeifer et al. 2005). This can be advantageous in that organizations become more creative and innovative and thus mobilize the collective intelligence to comply with the various requirements.

This means that despite their interesting content and structure, it is necessary to couple the excellence models with roadmaps to assist companies in their implementation. Moreover, (Zeyer 1996) states that operational practice showed that the success of a holistic management system not only depends on its operation-specific structures but is also crucially influenced by the type of implementation processes.

In this sense, (Bolboli and Reiche 2013), proposed a meta-model (Fig. 1.3) composed of three parts: Transformation Model, System Model, Implementation Model. It is designed and developed from a systemic perspective and describes both the fundamental elements to build a solid sustainable excellence system (System Model), and how to build it by describing the implementation steps (Transformation Model & Implementation Model) (Fig. 1.4).

The transformation model is represented as a control loop over four phases of change and four quality gates. HQs that are positioned after each phase are used to measure and evaluate the performance of each of the stages (Bolboli and Reiche 2013). The following activities (Table 1.3) describe the implementation mode:

Managerial pillar	Purpose
1- Management commitment	Ensuring the project's success
2- Involvement	Each employee must be aware of the company's objectives and must understand his or her position and role in achieving these objectives
3- Communication	Understand the how and why of a decision, Monitoring the achievement of objectives
4- Understanding	Understand what the problem is and where it occurs in order to define its scope
5- Measure	Define measurement indicators to quantify the problems and evaluate the performance obtained after the implementation of the solutions
6- Deployment	Translate the company's objectives into action on the ground, as part of a deployment plan that will be monitored by management
7-Implementation	Ensure the success of the project through the rigorous implementation of optimal solutions by the right people, To master the implemented solutions
8- Evaluation	Verify that identified problems have been resolved
9- Standardization	Standardize the method to perpetuate the results obtained following the resolution of the problems
10- Documentation	Capitalize on the know-how acquired and duplicate it elsewhere if necessary.

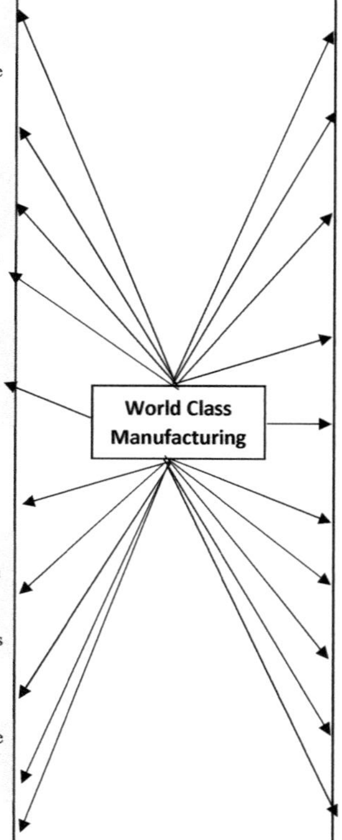

Technical pillar	Purpose
1- Safety, health and working environment	Reduce accidents in the working environment, Develop the culture and prevention, Improve working conditions and ergonomics at the workstation, Develop specific professional skills
2- Cost deployment,	Cost and loss analysis, Identify the main elements of losses in the production-logistics system, Quantify the potential and expected economic benefits, Allocate management commitment and resources to tasks with the greatest potential
3- Focused improvement,	Prioritize actions associated with the management of losses identified by cost deployment in order to increase the competitiveness of the product cost, Drastically reduce the most important losses, Eliminate non-value added activities, Develop specific professional problem solving skills.
4- Autonomous activities,	Combination of autonomous maintenance and work organization, improve the efficiency of the production system, Improve the working environment to prevent equipment degradation
5- Professional maintenance	Increase machine efficiency using failure mode analysis, Facilitate cooperation between operators and maintenance in order to achieve zero failure,
6- Quality control,	Reduce non-compliances, Increase culture quality and skills of employees
7- Logistics and customer service,	Reduce significantly the levels of stocks, Minimize the number of times the material is handled, from supplier to delivery
8- Early equipement management	Ensure fast and stable start-up of installations, reduce life cycle costs, Design systems easily maintained and inspected.
9- People development	Ensure, through a structured system of training, correct skills and abilities for each workstation.
10- Environnement	Comply with the requirements and standards of environmental management, Develop an energy culture and to reduce the energy costs and losses.

Fig. 1.3 The ten technical and managerial pillars of WCM

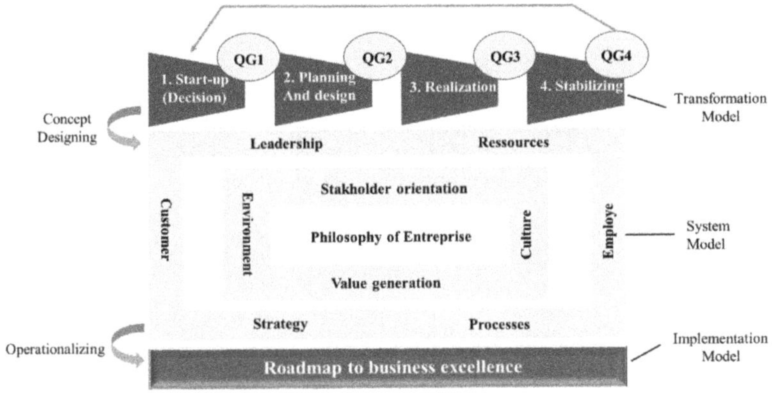

Fig. 1.4 Meta-model for business excellence (Bolboli and Reiche 2013)

Once the roadmap has been explained, it is important to draw attention to the notion of change management. Change is an inevitable consequence of the implementation of a project, in this case the continuous improvement approaches in the context of sustainable excellence. According to the authors (Robbins et al. 2014) organizational change is a change in the structure, technology, or personnel of an organization. Overall change management reflects the introduction of a new form of organization and production.

Indeed, the modifications and changes that accompany the implementation of projects, regardless of their nature, are felt through several aspects:

- **Organizational structure**: the adoption of certification approaches, for example, requires companies to structure themselves according to the process approach. This can lead to a modification of hierarchical relations (authorities and responsibilities), the redefinition of positions, the creation of coordination or supervision teams for projects on the scale of several processes (logistics, quality, production, etc.). Not to mention that all activities are arranged in a cross-functional approach facilitating coordination and collaboration.
- **Work methodologies:** for example, World Class Manufacturing implies the use of new methods and tools to accomplish activities. Also, the optimization of processes by Industry 4.0 technologies, the adoption of certain technologies or digital practices can affect the way things are done but also the results. For example, selling on the Internet has become commonplace today and is amplified during these times of health crisis.
- **Human resources**: the implementation of sustainable excellence approaches will certainly influence the corporate culture, notably through the dissemination and practice of new values and the adoption of new behaviors, for example, in the area of health and safety at work, the personnel must develop new attitudes towards professional risks. This implies that in order to build agile and flexible processes, a training plan is needed to sharpen the skills of the various employees.

Table 1.3 Transformation Model for Sustainable business excellence (Bolboli and Reiche 2013)

1. Stra-up (Decision)	2. Planning and design	3. Realization	4. Stabilizing
– Decision to implement; Economic analysis; – Establishment of a team made up of specialists from different areas of the enterprise; – Designation of promoters (in addition to the top-management) and people responsible for QGs; – General awareness (creating a sense of urgency); – Analysis of stakeholders and environmental spheres; – Creating values; and – Developing vision and mission	– Communicating vision and strategy; – Analyzing problems and identifying critical processes and activities to achieve the objectives; – Planning first successes; – Introducing plans and responsibilities in all areas; – Developing appropriate resolution measures; – Operational action plans for implementation of solution strategies – Developing of a catalog of individual measures in all areas; – Developing of a monitoring system regarding the BE activities; – Defining of milestones for regular review of target achievement; – Qualifying employees including specialized training to use the necessary tools and techniques for responsible people; and – Analyzing and discussing critical barrier and success factors for the implementation phase	(1)first stage, trial implementation of solution strategies for about six months; – Review of the program by trained auditors who are not involved in the activities that are to be checked; and – Determination of corrective actions (2) second stage of implementation regarding the corrective action up to one year; – Evaluation of the implementation activities and analysis of the successes and failures; and – Learning from successes and failures to identify the necessary activities of the last implementation stage (3) third stage of implementation up to six months; – Controlling and monitoring the results; and – Adjusting implementation plan to new conditions	– Continuing the review of goal achievement; – Integrating of BE activities into everyday work; – Examining environment; and – Experienced knowledge

It also appears that how change is "managed" has an impact on the success of the project (Hornstein 2015). For real change to occur, six parameters will need to be met (Forman et al. 2007) (Table 1.4).

In organizations, it has been widely reported that employees and the internal structure of the company resist change, especially when it is imposed (Préfontaine 2005).

Table 1.4 Parameter of success of a change according to (Forman et al. 2007)

Conditional parameters	Consequences in case of failure
Leadership	Doubt, loss of confidence
Vision and values	Confusion, loss of culture
Planning	Stagnation
Resources	Frustration, demotivation
Know-how (methods and tools)	Anxiety, stress and inefficiency
Recognition and reward	Passive resistance

This is a natural reaction, explain Aubert et al. (2010), which manifests itself within a social system to protect itself from a new situation perceived as threatening. Indeed, any change represents a source of ambiguity and uncertainty that can sometimes compromise the success of the project.

Fortunately, resistance to change is not irreversible (Yedder and Farhoud 2009). To lead change, the literature recommends approaches (Table 1.5) that revolve around communication, consultation and participation. Communication is crucial both to get employees to adhere to the project and to harmonize perceptions and clear up misunderstandings.

It should also be noted that new projects need change management actors. These actors, who play the role of federators and catalysts, are frequently endorsed by management and/or middle managers. In other words, these actors are called "change agents" defined by Robbins et al. (2014) as: "*Individuals or groups who serve as triggers and assume responsibilities for managing the change process*".

In practice, companies that decide to adopt an improvement approach very often call on external consultants and experts, which can represent both an advantage and a disadvantage. The advantage is that the external expert has the technical knowledge and the operational mastery of the improvement field (environmental certification, lean manufacturing…). Moreover, his intervention in several companies gives him a certain experience and a broad perception of the pitfalls to avoid. In addition, his fresh and objective look allows him to provide insightful and appropriate advice to achieve the recommended results. The authors, Robbins et al. (2014) draw our attention to the fact that an external consultant is sometimes synonymous with the possibility of misinterpretation of history, culture, operational procedures and personnel. Changes can also be more radical (for better or for worse) because the consultant does not have to deal with the repercussions. Therefore, they recommend the appointment of a process (or change) manager as it is more prudent, and he/she will be there to experience the consequences. From one point of view, the use of external consultants limits the company's ability to develop its own skills and collective intelligence. This has a direct impact on the collective and organizational learning that is strongly recommended in a sustainable excellence approach.

Table 1.5 Methods for dealing with resistance to change

Approach	Commonly used in situations	Advantages	Drawbacks
Education + communication	Where there is a lack of information or inaccurate information and analysis	Once persuaded, people will often help with the implementation of the change	Can be very time consuming if lots of people are involved
Participation + involvement	Where the initiators do not have all the information, they need to design the change, and where others have considerable power to resist	People who participate will be committed to implementing change, and any relevant information they have will be integrated into the change plan	Can be very time consuming if participators design an inappropriate change
Facilitation + support	Where people are resisting because of adjustment problems	No other approach works as well with adjustment problems	Can be time consuming, expensive, and still fail
Negotiation + agreement	Where someone or some group will clearly lose out in a change, and where that group has considerable power to resist	Sometimes it is a relatively easy way to avoid major resistance	Can be too expensive in many cases if it alerts others to negotiate for compliance
Manipulation + co-optation	Where other tactics will not work or are too expensive	It can be a relatively quick and inexpensive solution to resistance problems	Can lead to future problems if people feel manipulated
Explicit + implicit coercion	Where speed is essential, and the change initiators possess considerable power	It is speedy and can overcome any kind of resistance	Can be risky if it leaves people mad at the initiators

Source Kotter and Schlesinger (1979)

References

Allouche J, Laroche P (2005) Responsabilité sociale et performance financière des entreprise: une synthèse de la littérature. Colloque “Responsabilité sociale des entreprises: réalité, mythe ou mystification?”. Nancy-France

Amit R, Schoemaker PJ (1993) Strategic assets and organizational rent. Strateg Manag J 14:33–46

André J-M, Husser J, Barbat G, Lespinet-Najib V (2011) Le rapport de Développement Durable des enterprises Françaises: Quelles perspectives pour les parties pernantes? Revue Management & Avenir (N°48): 37–56

Aubert N, Jabes J, Laroche H, Enlart S, Gruère J-P (2010) Management, Aspects humains et organisationnels (éd. 9ème édition). PUF

BERGERY L, DEJOUX C (2006) Elargissement de la notion d’excellence: vers l’excellence durable. GestN 2000: Manag & Prospect 23(N°.6): 147–164

Bolboli SA, Reiche M (2013) A model for sustainable business excellence: implementation and the roadmap. The TQM Journal 25(4):331–346

Carmeli AT (2004) The relationships between intangible organizational elements and organizational performance. Strat Manag J 25(13): 1257–1278

Cho H-J, Pucik V (2005) Relationship between innovativeness, quality, growth

Dadfar H, Dahlgaard JJ, Afazeli S, Berge S (2015) Quality, export and domestic market performance: the case of pharmaceutical firms in Iran. Total Qual Manag Bus Excell 26(9–10):938–957

Dervitsiotis KN (2004) Navigating in turbulent environmental conditions for sustainable business. Total Qual Manag 15(N. 5–6): 807–827

DOU H (1995) Veille technologique et compétitivité: l'intelligence économique au service du développement industriel. Paris, Dunod

EFQM (2021) The EFQM Excellence Model. Consulté le Mars 20, 2021, sur https://www.efqm.org/

Elkington J (1997) Cannibals withforks The Triple bottom line of 21st Century business. Capstone Publishing Ltd., Oxford

Fernez-Walch S, Romon F (2006) Management de l'innovation: de la Stratégie aux projets. Vuibert, Paris

Forman B, Gey J-M, Bonnifet F (2007) Qualité Sécurité Environnement: Construire un système de management intégré. AFNOR.

Hennion R, Makhlouf A (2016) Les fiches Outils du Lean Six Sigma. Eyrolles, Paris

Heras-Saizarbitoria I, German-Arana L, Marti-Casadesus F (2006) A Delphi study on motivation for. Int J Qual & Reliab Manag 23(7): 807–827

Hornstein HA (2015) The integration of project management and organizational change management is now a necessity. Int J Proj Manag 33

ISO (2000) ISO 9004: quality management systems-guidelines for performance improvement. Deuxième Edition. ISO.

Jabnoun N (2019) A proposed model for sustainable. Manag Decis 58(2): 221–238

Ketchen D-J, Snow C-C, Street V-L (2004) Improving firm performance by matching strategic decision-making processes to competitive dynamics. Acad Manag Perspect 18(4):29–43

Koenig G (1994) Apprentissage Organisationnel: repérage des lieux. Revue Française de gestion: 83–95

Kotter J, Schlesinger LA (1979) Choosing strategies. Harv Bus Rev: 106–114

Lefebvre M-H (2018) Management de la Santé et de la Sécurité selon l'Iso 45001. AFNOR, Paris

Lesca H, Caron M-L (1995) Veille stratégique: créer une intelligence collective au sein de l'entreprise. REVUE FRANÇAISE GESTION

Löffler E (2001) Quality awards as a public sector benchmarking concept in OECD member countries: some guidelines for quality award organizers. Public Adm 21(1): 27–40

Mauléon F, Silva F (2009) Etats des lieux de la RSE et du développement durable en France. Revue Management & Avenir(N°.23): 23–35

Morin J (1985) L'excellence technologique. Publi Union, Paris

NGUYEN N (2006) Gestion de la qualité. Montréal, Chenelière Education

Peters T, Waterman R (1982) In search of excellence: lessons from America's best-run companies. (H. &. publishers, Éd.)

Pfeifer T, Schmitt R, Voigt T (2005) Managing change: quality-oriented design of strategic change processes. The TQM Magazine 17(4): 297–308

Préfontaine L (2005) Risk management in public and private partnership IT projects: an international study. Int Bus & Econ Res J 4(4)

Riztman L, Krajewski L, Renart J, Townley C (2013) Management des Opérations (éd. 2ème édition). Paris, Nouveaux Horizons

Robbins S, Decenzo D, Coulter M, Rüling C-C (2014) *Management: l'essentiel des concpets et pratiques*. (9th Éd.) France, Nouveaux Horizons

ROUBY E, SOLLE G (2001) Contribution à une définition du concept d'excellence organisationnelle. Université de Versailles, Saint Quentin, Colloque MAAOE

Sharma M, Kodali R (2008) TQM implementation elements for manufacturing excellence. The TQM Journal 20(6): 599–621

Villain J (1990) L’entreprise aux aguets: information, surveillance de l’environnement, propriété et protection industrielles, espionnage et contre-espionnage au service de la compétitivité. Masson, Paris

Yamashina H (1995) Japanese manufacturing strategy and the role of toal productive maintenance. J Qual Maint Eng 1:27–38

Yedder M-B, Farhoud M (2009) Le développement durable est-il bienvenu dans les organisations ? Cas de l’implantation d’un Système de Management Environnemental en Tunisie. Développement durable & territoire

Zeyer U (1996) Implementierungsmanagement : ein konzeptioneller Ansatz am Beispiel der Implementierung von Lean Management. Hampp. Journal. Emerald Group Publishing Limited, München

Chapter 2
Sustainable Business Excellence in SMEs: Opportunities and Challenges

Objectives

1. Provide an overview of the sustainable business excellence
2. Provide SMEs with a conceptual model for the integration of sustainable business excellence components

2.1 Introduction

The past literature alongside the most common practitioner-oriented conceptual frameworks of business excellence (BE) have historically been mainly developed and adapted for large companies (Sternad et al. 2019; Belhadi et al. 2016). However, recent studies have clearly highlighted the tremendous role of Small and Medium-sized Enterprises (SMEs) as a driving force behind many of the world's leading economies (Belhadi et al. 2018a). According to the Organization for Economic Co-operation and Development (OECD), SMEs account of more than 99% of all business around the world (Marchese et al. 2019). These companies contributed by more than 60% of the increase in employment in the USA in the period between mid-2009 and mid-2013 and more than 70% of jobs creation in the EU in the year 2014 (Sternad et al. 2019). In spite of their vital role in the economic growth, the productivity of SMEs, in general, is shown to be low compared to their large counterparts. The OECD stated that the productivity of manufacturing-based small and medium enterprises relative to large companies was respectively 62% and 75% in 2014 (Marchese et al. 2019). Further, it has been established that SMEs in the manufacturing sector are responsible for a large portion of the world's consumption of resources, air and water pollution, and waste generation (Jabbour et al. 2020). Although the individual contribution of each company is low, the cumulative damage and footprint that SMEs cause to the environment is significant (Afum et al. 2020). This suggests that SMEs are still lagging behind larger businesses in the adoption of BE frameworks especially with the strong evidence that these frameworks can benefit companies of all types

F. E. Touriki et al., *Sustainable Excellence in Small and Medium Sized Enterprises*, Industrial Ecology, https://doi.org/10.1007/978-981-19-0371-7_2

and sizes in terms of improving their sustainable performance including economic, environmental and societal aspects (Belhadi et al. 2018b; Cherrafi et al. 2017).

The most widely used BE models are suffering from two main shortcomings, hindering them from being totally adopted to enhance sustainable performance of SMEs. First, they are particularly adapted to the requirements of larger companies. For instance, models based on lean production, six sigma or total quality management was originally devised by large firms such as Toyota or General Electrics (Sternad et al. 2019). Therefore, these models seem to present a holistic nature and their implementation is relatively complex and more difficult to manage for SMEs (Panizzolo et al. 2012; Achanga et al. 2006). Second, the current models are not totally confirmed to bring the needed improvement in sustainable performance. According to Choudhary et al. (2019), the integration of approaches such as lean and six sigma with sustainable performance does not seem to be obvious and several adaptations are needed. These adaptations are yet to be achieved in the literature (Siegel et al. 2019). Despite the recent attempts to develop versions of these models that are more suitable to the needs of sustainability in SMEs (Choudhary et al. 2019; Siegel et al. 2019; Belhadi et al. 2018b), no widely used approach that is specifically adapted to the needs of sustainability in SMEs has yet emerged.

SMEs are actually under growing pressure to manage their operations in a responsible manner considering economic, environmental and social performance. This situation raises several questions:

1. What are the main sustainable opportunities offered for SMEs through implementing BE approaches?
2. What are the motives and obstacles that SME managers see for adopting a sustainable BE approach in their companies?
3. What are the drivers and enablers for SMEs to be implemented as an entry-level approach to sustainable BE?

The answers to these research questions can help to better understand how SMEs can be encouraged to consider the use of BE models for sustainable performance, and how these models need to be adapted and interact with different enablers to suit the specific needs of smaller businesses.

2.2 Sustainable Business Excellence and SMEs: Overview

2.2.1 From Operational Performance to Sustainable Performance

Traditionally, organizations are used to seek for continuous improvement in specific workshop measures such as quality, cost and delay (Belhadi et al. 2016; Sternad et al. 2019). However, managers of manufacturing firms, recently, have realized that the organizational survival does not require only these lower-level measures in the long-term but equally needs a substantial strive to enhance the performance at multiple other aspects (Belhadi et al. 2018b). Obviously, an organization's performance should be evaluated according to the results it achieves. In fact, the firms should be demonstrating sustainable results on many levels such as results related to customers, financial and market, people, processes and the external environment (Caiado et al. 2019). This integrated approach of performance and results should be used to assess the level to which the firm has progressed to attain its vision and mission (Jabbour et al. 2020). The excellent firms strive to reach superior achievements that are sustainable, comprehensive and in harmony with the requirement of all stakeholders (Mangla et al. 2020). This is a natural outcome of the strategy, policies, practices and course plans that are implemented regularly in all departments of the organization. These firms are characterized by the importance of balancing internal and external requirements for performance. High performing organizations are poised for market leadership and growth. They achieve outstanding financial and customer achievements while meeting the interests of employees and other stakeholders. They balance short-term results and long-term outcomes. Generating value for internal and external stakeholders enables the firm to generate profit, trust and to contribute to the community (Jankalová and Jankal 2020).

For all these reasons, the sustainable performance comes to the front as a concept of that concurrently incorporates environmental, economic, and societal measures to assess and achieve business performance (Cherrafi et al. 2017). The Triple Bottom Line (TBL) concept clubs all these three metrics of sustainability and integrates the requirement of internal and external stakeholders in business performance policies (Mangla et al. 2020). Yet, several authors highlighted the urgent need to swift from operational performance of the firms to a wider sphere that needs to be evaluated based on the firms' trade-offs with TBL model of sustainability (Kamble et al. 2020; Belhadi et al. 2020; Gomes et al. 2019).

2.2.2 *Sustainable Business Excellence*

In a fast-changing context, organizational survival relies not only on operating in the most profitable, effective and efficient way possible, but also on its compromise with environmental regulations and social requirement and the adoption of evolving strategies (Afum et al. 2020). Hence, sustainability in operations is emerging as a relevant challenge for organizations (Kamble et al. 2018). BE programs such as quality management, lean manufacturing, six sigma, Industry 4.0 are frequently studied from sustainable performance perspectives (Belhadi et al. 2020; Kamble et al. 2020; Belhadi et al. 2018b; Cherrafi et al. 2017). From this evolves the so-called 'Sustainable Business Excellence' (SBE), which delineates the interrelation between BE strategies and sustainable performance. Table 2.1 summarizes the BE approaches and their reported effect on sustainable performance according to the past literature.

Table 2.1 BE approaches and their reported effect on sustainable performance

BE approaches	Definition	Effect on sustainable performance	References
Lean manufacturing	Primarily introduced by Taiichi Ohno at the Toyota Motor Company in the 1950s to reduce wastes, achieve an involving approach associating employees, suppliers and customers	**Operational benefits:** • Lead time reduction • Inventory reductions • Cost saving	Belhadi et al. (2018a), Kamble et al. (2020), Bai et al. (2019), and Dombrowski et al. (2010)
		Environmental benefits: • Greenhouse gases emissions reduction • Energy consumption reduction	Belhadi et al. (2018b), Mangla et al. (2020), Choudhary et al. (2019), Caiado et al. (2019), Carvalho et al. (2017), and Bai et al. (2019)
		Social benefits: • Workplace safety • Mental health enhancement	Henao et al. (2019), Distelhorst et al. (2017), and Signoretti (2019)
Six Sigma	Six Sigma was created by Bill Smith at Motorola Corporation, in the 1980s, seeking to reduce errors and defects	**Economic benefits:** • Quality defect reduction • Cost saving	
		Environmental benefits: • Water and material consumption reduction • Energy consumption reduction	
		Social benefits: • Workplace safety • Mental health enhancement	Sony et al. (2020)

Lean manufacturing (LM) is an integrated approaches coined by Taiichi Ohno with aim of targeting and eradicating wastes all over the organizations (Belhadi et al. 2018b). The concept of lean is based on considering all activities consuming resources without creating value for the final client as wasteful and need to be eradicated (Zhou 2012; Shah and Ward 2007). The central goal of lean is then to eradicate non-value-added operations everywhere extending from suppliers to the end costumer. In the literature, several methodologies based on popular lean tools have showed relevant outcomes in sustainable performance enhancement.

As expected, the relationship between LM and Operational Performance (OP) was the first to be studied, in investigations dating back to 1980s. According to Shah and Ward (2007), successful implementation of lean practices can lead to an OP increase of up to 23%, when evaluating workshops aspects such as quality, lead-time, unit cost, and flexibility. Büyüközkan et al. (2015) concluded that the full adoption of LM tends to affect OP metrics by up to 16%. Performance can even increase up to 20% when financial variables are considered in the performance measure. However, it can also drop to just 13% with the average level of LM implementation found in companies (Henao et al. 2019). Furthermore, most empirical research backs the proposition that LM is positively related to Environmental Performance (EP). Choudhary et al. (2019) supported that LM can critically influence EP. Other studies (Bai et al. 2019; Belhadi et al. 2018b; Carvalho et al. 2017) postulate that LM shows similar positive effects on both OP and EP. The continuous effort through LM to reduce operational waste either from discarded materials, consumption of energy or water usage typically means lower pollutant emissions, supporting improved environmental performance (Belhadi et al. 2018b). It has been established that LM generates energy and water efficiency due to the core principle of zero waste (Choudhary et al. 2019). On the social side, authors such as Henao et al. (2019) suggested that LM could minimize risks on employees due to the reduction of poor health and safety conditions.

When workers are satisfied and comfortable in the working conditions in a lean environment, it would result in sustainable lean performance enhancement in the long-term. R. Sharma (2012) presented a framework illustrating how LM practices integrated with ergonomic interventions can help in achieving business excellence in terms of cost effectiveness, faster delivery and quality improvement while improving the work conditions. Botti, Mora, and Regattieri (2017) developed a mathematical model to design a lean hybrid assembly line by incorporating the physical factor into lean manufacturing to enhance performance for sustainable outcomes.

Workers will experience increased autonomy and involvement in decision making, primarily through their participation in problem-solving activities (e.g. Womack et al. 1990).

On the other hand, the pursuit of standardization through LM, without the proper level of employee empowerment, can end in routine operation, which is often associated with a lack of improvement, slippage back to old ways of working, laxity, rule breakage, defiance, and even sabotage (Ketokivi and Schroeder 2004). In addition, if employees feel that the management does not appreciate their efforts, they may be discouraged, and the lean manufacturing effort will fail. Though it is often best to drive change from the factory floor (Gemba), show visible management connected to the project, and participate in the LM events. Consequently, we can safely conclude that LM practices impact positively the social aspect of sustainability if certain conditions are met.

Increased popularity and interest of LSS is noticed within small and medium sized enterprises. The term Lean six sigma was first introduced in the literature in 2000s (Timans et al. 2012) combining both lean and six sigma and overcoming the shortcomings of both (cherrafi et al. 2016).

It is defined as a business strategy and methodology that aims to increase the overall performance, improve customer satisfaction, leadership and bottom-line results by acting on quality, delay and costs (Snee 2010) and helping to prevent product defects through the elimination of waste and process variations.

Authors have largely elaborated on the benefits of LSS on both operational and environmental performance. Fatemi and Franchetti (2016) stated the existence of improvements in cost and environmental criteria through the application of LSS. R.D. Snee (2010) indicated that companies noticed an increase in the job effectiveness when deploying LSS with a return of 1–2% of sales per year for large companies and 3–4% of sales per year for medium size ones, which can be translated to significant savings.

In a case study conducted by Furukawa et al. (2016), results of pre and post implementation of LSS showed a reduction in packaging and plastic bags, in addition to the cost reduction of material disposals. Pollution prevention programs were improved, environmental impacts of discharging the produced water in oil fields were reduced, after the implementation of the six sigma projects.

On the social side, M. Sony, S. Naik, and J. Antony (2020) determined that LSS has an impact on decent work, the health and well-being of society, public services and infrastructure, quality education for society, and climate action.

The employee's satisfaction is often related to changes within the personal and new skill development, it is increased when employees feel more recognition from management due to the LSS efforts (Sony et al. 2020), however the impact of LSS on employee training is very significant. Six Sigma in particular requires a high degree of training and effort by employees consequently in order for LSS to succeed, organizations need to train employees at all levels (Kumar et al. 2006). It is also noted that after the implementation of LSS there was a drastic reduction in the rejection rate. As a result, workers sense a high degree of accomplishment.

2.3 Small and Medium-Sized Enterprises: Overview

SMEs are considered as the efficient motor of every market economy. On the other hand, they are much more different from large enterprises and many methodologies for lean implementation cannot be adopted by SMEs due to several issues (Dombrowski et al. 2010; Yusof and Aspinwall 2000). Therefore, a prerequisite to scientifically study the implementation of lean in SME is to delineate the SME from large enterprise and to pave the way for assigning a clear definition for this specific type of companies (Loecher 2000).

2.3.1 Definition of SMEs

Overall, SME is a category of companies from all sectors which does not exceed a given set of size thresholds (Loecher 2000). However, there is no consensus on a given size threshold between countries and industries (Deros et al. 2006). Accordingly, SMEs are defined by a different set quantitative indicator, like turnover, total capital, balance-sheet total, invested capital, shares, marketplace, sales volume or assets value and number of workforces which widely vary depending on the country and even the type of industry (Deros et al. 2006; Loecher 2000). Table 2.2 illustrates a recap of SMEs definitions in the manufacturing sector of given economies.

It is observed that SMEs are defined most often by numbers of employees, turnover and investment or all of them. This is true for both more developed countries (European Union, US, and Japan) and less developed countries (China, India, Malaysia and Morocco). Even though turnover and investment are part of the definition criteria in most countries, the number of employees is appeared to be the overriding criterion adopted to delineate the boundaries between SMEs and large enterprises (Deros et al. 2006). Therefore, in order to conform with the same set of criteria used in previous studies in our research field to identify SMEs, we consider SMEs as companies with fewer than 250 employees (Zhou 2012; Timans et al. 2012; Panizzolo et al. 2012; Achanga et al. 2006; Deros et al. 2006).

Table 2.2 Definition of SMEs in the different sectors of selected countries (more and less developed countries)

Category	Country	Sector	Criteria
More developed countries	European Union	Manufacturing	< 250 employees < 50 M € turnover or < 43 M € Balance sheet total
	US	Manufacturing	< 1 500 employees < 17 M $ turnover
		Commercial	< 500 employees < 13.5 M $ de turnover
		Service	< 500 employees < 5 M $ turnover
	Japan	Manufacturing and other	< 300 employees < 300 M ¥ fixed assets
		Wholesale	< 100 M ¥ fixed assets
		Retail	< 50 M ¥ fixed assets
		Service	< 50 M ¥ fixed assets
Less developed countries	China	Varies with industry	Usually < 300 employees
	India	Manufacturing	Up to Rs10.00 million in plant and machinery
	Malaysia	Manufacturing	< 175 full time workers Investment US$1 million
	Morocco	Manufacturing	< 200 employees < 75 M Dh turnover or < 50 M Dh Balance sheet total

2.3.2 Characteristics of Small and Medium-Sized Enterprises

Although many large companies started out as SMEs with very small capitals like Toyota, Ford Motors, Hewlett-Packard and Microsoft, SMEs could not be considered as a thumbnail of a large company (Deros et al. 2006). Indeed, SMEs differ largely from large companies regarding strategies, structures and ways of doing business (Welsh and White 1981). Thus, studying the main characteristics that distinguish SMEs from large organizations is very helpful. Table 2.3 depicts the principal characteristics of SMEs versus large enterprises divided into four categories.

As for policy and procedures, Ghobadian and Gallear (1996) argue that SMEs are characterized by a flat structure and closely linked to the manager/ owner and less interfaces between departments. This creates an environment flexible and conducive for change. Large organizations are on the other hand usually bureaucratic with several layers of management. A majority of SMEs rely on a short-term vision that hinders in general the continuous improvement unlike large organizations who promote usually the continuous improvement through their Medium and long-term vision. In turn, SMEs are more advantageous concerning flexible organization and flows since it fosters simple communication and short decision-making chains (Deros et al. 2006; Ghobadian and Gallear 1996).

In terms of the way to utilize the company's assets and resources, SMEs are characterized by a low degree of standardization and formalization promoted by the lack of specified training. SMEs, on the other hand, are in a more advantageous position since they rely on a higher level of the use of standards and procedures (Ghobadian and Gallear 1996). Also, large organizations possess a well-trained human capital, sufficient resources and expertise and developed infrastructure unlike SMEs whose human and material resources are modest and generally inadequate.

The prevailing culture and behaviors in SMEs differ from that in large enterprises since in SMEs there is no departmental or functional culture which inhibits the creation of interest groups and then establishes a corporate mind-set more likely results oriented and conducive for new change initiatives. By contrast, in large enterprises a rigid corporate culture is dominating operations and behaviors, which is less conducive to new change initiatives (Loecher 2000).

Finally, concerning the market positioning, SMEs have not a high influence over their suppliers, which are most often large enterprises. Their products and services are mostly destined for local market unlike large companies who can easily penetrate national and international markets. However, the simple structure of SMEs makes them more responsive to the market change than their large counterparts.

Table 2.3 Comparison between the characteristics of large enterprises and SMEs

	Large enterprises	SMEs
Policy and procedures	Bureaucratic with several layers of management	Flat with very few layers of management
	Clear and extensive functional division of activities. High degree of specialization	Division of activities limited and unclear. Low degree of specialization
	Formal evaluation, control and reporting procedures	Informal evaluation, control and reporting procedures
	Medium- and long-term vision that promote the continuous improvement	Short term vision that hinder the continuous improvement
	Rigid organization and flows	Flexible organization and flows
	Extended decision-making chain	Short decision-making chain
	Tenuous link between company and owner	Close tie between company and owner
Utilization of assets and resources	Ample human capital, financial resources and know-how	Modest human capital, financial resources and know-how
	Company manager is responsible primarily for business/ organizational duties	Company manager personally works predominantly in the area of production or in the technical department
	Generally, developed infrastructure	Inadequate infrastructure
	Specified training budget	No specified training budget
	High degree of standardization and formalization	Low degree of standardization and formalization
Culture and behavior	High degree of resistance to change	Negligible resistance to change
	Many interest groups	Very few interest groups
	Strong departmental/functional mind set	Absence of departmental/functional mind set. Corporate mind-set
	Control oriented	Result oriented
	Rigid corporate culture dominating operations and behaviors	Operations and behavior of employees influenced by owners'/managers" ethos and outlook
	Corporate mind-set is not conducive for new change initiatives	Corporate mind-set is conducive for new change initiatives
Market positioning	Normally slow response to the market changes	Normally rapid response to the market changes

(continued)

Table 2.3 (continued)

	Large enterprises	SMEs
	High influence over their suppliers	Low influence over their suppliers
	Product and services mostly for national and international market	Product and services mostly for local market
	Extensive external contacts	Limited external contacts

2.4 Sustainable Business Excellence in SMEs

2.4.1 Opportunities of Sustainable Business Excellence in SMEs

Past literature has widely investigated why organizations implement sustainable business excellence approaches such as LM and LSS. However, only few studies have considered particularly the opportunities for adopting SBE approaches in SMEs. Table 2.4 presents a review of the available literature on the opportunities of SBE approaches in the context of SMEs.

Besides their proven benefits in the three dimension of sustainability i.e., economic, social and environmental aspects, the opportunities provided to SMEs by adopting a SBE approach are usually categorized into internal and external. Internal opportunities can, for example, include the desire to improve the SME's sustainability on economic side such as, quality, productivity, delay and cost (Belhadi et al. 2018a; Dombrowski et al. 2010). This can include also the environmental performance such as emission reduction and resource efficiency and the social performance such as workplace safety and mental health environment (Belhadi et al. 2020; Henao et al. 2019). Moreover, SMEs could implement SBE approach in order to level up their organization to the world-class level (Sternad et al. 2019). According to authors such

Table 2.4 Opportunities of SBE approaches for SMEs

Categories	Opportunities	References
Internal	Sustainable profitability	Belhadi et al. (2020), Choudhary et al. (2019), Jabbour et al. (2020), and Panizzolo et al. (2012)
	Competitiveness and World class	Sternad et al. (2019), Zhou (2012), and Timans et al. (2012)
	Overall strategic framework	Belhadi et al. (2016) and Cherrafi et al. (2017)
External	External market reasons	Sternad et al. (2019), Marchese et al. (2019), and Siegel et al. (2019)
	External reasons of requirements	Sternad et al. (2019) and Belhadi et al. (2019)

as Belhadi et al. (2016) and Cherrafi et al. (2017) SBE approaches such as LM and LSS could constitute an overall strategic framework for the steering of the SME activity. Besides, several authors observed that SBE could provide several external opportunities. We follow Sternad et al. (2019) to classify the external opportunities into 'external market reasons' (such as improving the firm reputation and the competitive position or showing the effectiveness of management practices) and 'external reasons of requirements' (such as customer demands or general requirements for competing in a specific sector).

2.4.2 Challenges of Sustainable Excellence in SMEs

Several studies in the literature have addressed the challenges and barriers that hinder SMEs from fully benefiting from the different SBE approaches. In fact, achieving sustainable business excellence in SMEs is not easy when we consider many key technical, human, financial, organizational and external challenges (Belhadi et al. 2017).

- **Technical challenges**

  Usually, SMEs face technical issues when dealing with transformative and business excellence approaches. Given lack of skilled resources and funding, SMEs generally have limited access to new and advanced technologies in terms of processes, tools and software (Choudhary et al. 2019). Further, the poor infrastructure facilities including power, water and equipment generates huge sustainability-related wastes and make the implementation of SBE very difficult (Siegel et al. 2019).

- **Human related challenges**

  In SMEs, each person has a key duty, most often numerous, leading to little spare resources (Siegel et al. 2019). Therefore, dedicating specific employees for business excellence project will leave them less time for their actual work. This hinders SMEs' management from engaging in training their employees, which in turn lead to unqualified people. That is why a workforce is considered a critical failure factor for SMEs (Achanga et al. 2006).

- **Financial challenges**

  One challenge that is commonly mentioned in the literature is that SMEs are most certainly struggling against a lack of resources. According to the OECD reports, most problems of SMEs have a financial nature. SMEs have limited access to capital markets in addition to a lack of realistic planning and financial evaluation procedures.

  As a consequence, SMEs cannot afford to invest in new and advanced technologies and engage in the SBE implementation.

  The lack of skilled resources within SMEs, as mentioned before, would acquire external and technical assistance and expertise leading to additional costs (Belhadi et al. 2018a; Jabbour et al. 2020).

- **Organizational challenges**

Inadequate management capacity and resistance to change were cited as the two main organizational challenges facing SMEs, as managers and business owners are convinced that changing the current production process and technologies would result in potential risks. Negative values in companies, relentless beliefs of employees, and resistance to change or adoption of new technologies are owing to the fear of complexity in the use of the latest technology and the lack of confidence among the staff and managers to employ new technologies to improve business performance [Role of Social and Technological Challenges in Achieving a Sustainable Competitive Advantage and Sustainable Business Performance.

While basic issues of SMEs are fundamentally reported as the lack of new technologies used, insufficient business-related training...inadequate managerial capabilities is also stated in the list. It was determined that SMEs managers had access to fewer strategic tools than other managers. This can hinder the implementation of the SBE projects given the fact that managers can be ill-qualified to handle the nature of such projects.

- **External challenges**

There is an increasing pressure on SMEs from the supplier's side in order to meet their specifications and fit into their requirements category, moving towards just-in-time inventory and manufacturing requirements, thus SMEs are highly pressured to meet these expectations.

Further, trade arrangements such as NAFTA (North American Free Trade Agreement), CAFTA-DR (Central American-Dominican Republic Free Trade Agreement), EU (the European Union), and APEC (Asia–Pacific Economic Cooperation), AELE (European Free Trade Association)..., exchange rate fluctuations and change in government requirements are components of the external regulatory challenges that SMEs need to overcome to adapt and change to achieve Sustainable Competitive Advantage.

Special attention needs to be given to the demographic changes associated with the aging workforce and the declining entrant pool, as these challenges are not always visible to small business owners consequently such a major shift in the labor market can be blind siding (October 2005). This can worsen even more the Human related challenges mentioned earlier given the fact that the multi-tasking responsibilities of small business owners and managers implies that they don't have enough time to tackle these challenges and be well-informed about them.

2.5 Conceptual Model for BSE in SMEs

The study of successful LM and LSS initiatives on the three pillars of sustainability within SMEs presented in Fig. 2.1 revealed plenty of elements to be capitalized as an integral framework of the implementation of BSE in SMEs. The proposed

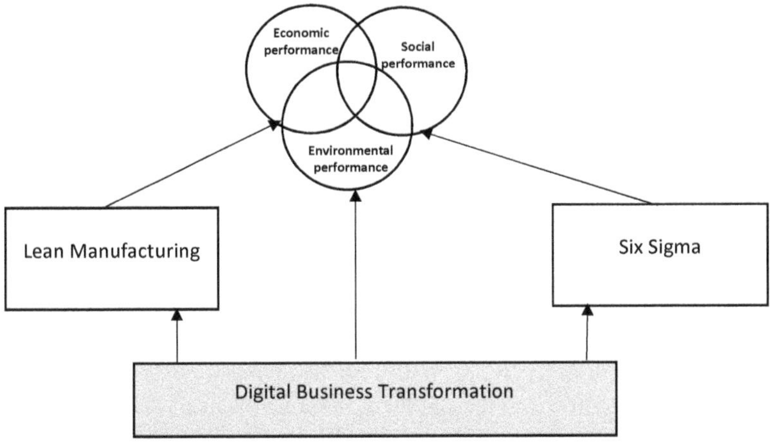

Fig. 2.1 Conceptual framework of SBE in SMEs

framework considers digital business transformation as the driven force that enables and supports the SBE elements. While also stressing on the direct impact of LM and LSS on sustainability.

New era of lean manufacturing is being introduced thanks to the digital technologies. Sanders, Elangeswaran, and Wulfsberg (2016) claimed that the barriers of LM can be overcome through the industry 4.0 applications, these technologies can help industries to become lean without the need to maintain conscious and persistent efforts concluding a positive correlation between lean manufacturing and Industry 4.0.

Six sigma as a data driven approach can be enhanced by the power of advanced analytics, resulting in time reduction that is usually high in term of data collection, and in a faster treatment and processing in order to highlight the root causes of variations in the most efficient way (Arcidiacono and Pieroni 2018). This in order to finally make effective decisions for quality problems by benefiting from advanced techniques such as big data analytics and process mining in addition to traditional techniques (Dogan et al. 2018, July).

Industry 4.0 is assumed to be a new business approach that can help organisations and society to move towards sustainable development (de Sousa Jabbour et al. 2018) through the integration of not only environmental protection, control initiatives but also process safety, such as resources efficiency, employee and community welfare, smarter and flexible processes (Luthra and Mangla 2018). Its initiatives transform a production system and supply chain into a smart system based on CPS, making the manufacturing system more flexible, economical, and environmentally friendly.

The proposed framework is self-explanatory and easy to understand for SMEs with a simple structure, also the links between the different elements of the framework are clearly distinguished, if followed efficiently SMEs may harvest its benefits and achieve sustainable business excellence. The role of training at all levels of the organization is highly stressed at this stage in order to succeed in the implementation of this framework, as elaborated in the literature well trained and motivated staff with the right skills and expertise can make or break the SBE projects. In sum, the proposed framework is suitable and applicable to SMEs in different industries, since it provides a practical methodology that summarizes the main assets for the introduction of SBE in a unique type of businesses such as SMEs.

2.6 Conclusion

This paper aims to elaborate on the implementation of the SBE on SMEs. This rising economic force accounts for more than 90% of all businesses around the world (Marchese et al. 2019), however there is little literature on sustainable business excellence within SMEs. Main sustainable opportunities offered for SMEs through implementing BE approaches were identified, in addition to the motives that encourage organizations to adopt a sustainable BE approach along with the obstacles that can hinder this process. Furthermore, drivers and enablers for SMEs were elaborated in order to implement SBE projects.

The added value of the paper is a conceptual framework proposition that has a simple structure including four main elements. The digital business transformation acting as the base that enables the SBE through the direct impact on LM, LSS, sustainability and its three pillars. LM and LSS are considered as two supporting elements that can positively affect sustainability if certain conditions are met.

In a nutshell this work can motivate managers and practitioners of SMEs to consider the adoption of SBEs through understanding its benefits, opportunities and prevailing challenges by the implementation of DBT. However, and to further improve its fulfillment the authors recommend a validation of the proposed framework by carrying out an implementation in a typical organization. Moreover, the framework could be enriched by studying other experiences like case studies or even surveys.

References

Achanga P, Shehab E, Roy R, Nelder G (2006) Critical success factors for lean implementation within SMEs. J Manuf Technol Manag 17(4):460–471

Afum E et al (2020) Green manufacturing practices and sustainable performance among Ghanaian manufacturing SMEs: the explanatory link of green supply chain integration. MEQ 31(6):1457–1475

Alavi Yacoubi L, Tourabi A (2020) La relation: ambidextrie entrepreneuriale et résilience organisationnelle des PME. Rev Int Sci Gest 3(2):365–388

Arcidiacono G, Pieroni A (2018) The revolution lean six sigma 4.0. Int J Adv Sci Eng Inf Technol 8(1):141–149

Bai C, Satir A, Sarkis J (2019) Investing in lean manufacturing practices: an environmental and operational perspective. Int J Prod Res 57(4):1037–1051

Belhadi A et al (2020) The integrated effect of Big Data Analytics, Lean Six Sigma and Green Manufacturing on the environmental performance of manufacturing companies: The case of North Africa. J Clean Prod 252:119903
Belhadi A, Touriki FE, Elfezazi S (2016) A framework for effective implementation of lean production in small and medium-sized enterprises. J Ind Eng Manag 9(3):786–810
Belhadi A, Touriki FE, Elfezazi S (2017) Prioritizing the solutions of lean implementation in SMEs to overcome its barriers. J Manuf Technol Manag 28(8):1115–1139
Belhadi A, Touriki FE, ElFezazi S (2018a) Lean implementation in small and medium-sized enterprises in less developed countries: Some empirical evidences from North Africa. J Small Bus Manage 56(sup1):132–153
Belhadi A, Touriki FE, Elfezazi S (2019) Evaluation of critical success factors (CSFs) to lean implementation in SMEs using AHP: A case study. Int J Lean Six Sigma 10(3):803–829
Belhadi A, Touriki FE, Fezazi SE (2018b) Benefits of adopting lean production on green performance of SMEs: a case study. Prod Plan Control 29(11):873–894
Belhadi A et al (2019) Understanding the capabilities of Big Data Analytics for manufacturing process: insights from literature review and multiple case study. Comput Ind Eng 137:106099
Botti L, Mora C, Regattieri A (2017) Integrating ergonomics and lean manufacturing principles in a hybrid assembly line. Comput Ind Eng 111:481–491
Büyüközkan G, Kayakutlu G, Karakadılar İS (2015) Assessment of lean manufacturing effect on business performance using Bayesian Belief Networks. Expert Syst Appl 42(19):6539–6551
Caiado RGG et al (2019) Towards sustainability by aligning operational programmes and sustainable performance measures. Prod Plan Control 30(5–6):413–425
Carvalho H, Govindan K, Azevedo SG, Cruz-Machado V (2017) Modelling green and lean supply chains: an eco-efficiency perspective. Resour Conserv Recycl 120:75–87
Cherrafi A, Elfezazi S, Chiarini A, Mokhlis A, Benhida K (2016) The integration of lean manufacturing, Six Sigma and sustainability: A literature review and future research directions for developing a specific model. J Clean Prod 139:828–846
Cherrafi A et al (2017) A framework for the integration of Green and Lean Six Sigma for superior sustainability performance. Int J Prod Res 55(15):4481–4515
Choudhary S et al (2019) An integrated lean and green approach for improving sustainability performance: a case study of a packaging manufacturing SME in the U.K. Prod Plan Control 30(5–6):353–368
Deros BM, Yusof SM, Salleh AM (2006) A benchmarking implementation framework for automotive manufacturing SMEs. Benchmarking: An International Journal 13(4):396–430
Distelhorst G, Hainmueller J, Locke RM (2017) Does lean improve labor standards? Management and social performance in the Nike supply chain. Manage Sci 63(3):707–728
Dogan O, Gurcan OF (2018, July) Data perspective of Lean Six Sigma in industry 4.0 Era: A guide to improve quality. In Proceedings of the international conference on industrial engineering and operations management. Paris
Dombrowski U, Crespo I, Zahn T (2010) Adaptive configuration of a lean production system in small and medium-sized enterprises. Prod Eng Res Devel 4(4):341–348
Dowlatshahi S, Taham F (2009) The development of a conceptual framework for just-in-time implementation in SMEs. Prod Plan Control 20(7):611–621
Fatemi S, Franchetti MJ (2016) An application of sustainable lean and green strategy with a Six Sigma approach on a manufacturing system. Int J Six Sigma Compet Advant 10(1):62–75
Furukawa PDO, Cunha ICKO, Pedreira MDLG, Marck PB (2016) Environmental sustainability in medication processes performed in hospital nursing care. Acta Paulista de Enfermagem 29:316–324
Garza-Reyes JA, Villarreal B, Kumar V, Ruiz PM (2016) Lean and green in the transport and logistics sector – A case study of simultaneous deployment. Prod Plan Control 27(15):1221–1232
Ghobadian A, Gallear D (1996) Total quality management in SMEs. Omega 24(1):83–106

Gnanaraj SM, Devadasan SR, Shalij PR (2010) Current state maps on the implementation of lean and Six-Sigma paradigms and an exclusive model for deploying Lean Six-Sigma in SMEs. Int J Product Qual Manag 5(3):286–309

Gomes PJ, Silva GM, Sarkis J (2019) Exploring the relationship between quality ambidexterity and sustainable production. Int J Prod Econ 224:107560

Henao R, Sarache W, Gómez I (2019) Lean manufacturing and sustainable performance: Trends and future challenges. J Clean Prod 208:99–116

Jabbour ABLdS, Ndubisi NO, Seles BMRP (2020) Sustainable development in Asian manufacturing SMEs: Progress and directions. Int J Prod Econ 225:107567

Jankalová M, Jankal R (2020) How to characterize business excellence and determine the relation between business excellence and sustainability. Sustainability 12:6198

Kamble S, Gunasekaran A, Dhone NC (2020) Industry 4.0 and lean manufacturing practices for sustainable organisational performance in Indian manufacturing companies. Int J Prod Res 58(5):1319–1337

Kamble SS, Gunasekaran A, Gawankar SA (2018) Sustainable Industry 4.0 framework: A systematic literature review identifying the current trends and future perspectives. Process Saf Environ Prot 117:408–425

Karlsson C, Åhlström P (1996) Assessing changes towards lean production. Int J Oper Prod Manag 16(2):24–41

Ketokivi M, Schroeder R (2004) Manufacturing practices, strategic fit and performance: a routine-based view. Int J Oper Prod Manage

Kumar et al (2006) Cité dans Chapitre 1/Tableau 13: Summary of Lean applications in SMEs for selected countries

Loecher U (2000) Small and medium-sized enterprises—delimitation and the European definition in the area of industrial business. Eur Bus Rev 12(5):261–264

Lopes de Sousa Jabbour AB, Jabbour CJC, Godinho Filho M, Roubaud D (2018) Industry 4.0 and the circular economy: A proposed research agenda and original roadmap for sustainable operations. Ann Oper Res 270(1):273–286

Mangla SK et al (2020) Operational excellence for improving sustainable supply chain performance. Resour Conserv Recycl 162:105025

Marchese M, Giuliani E, Salazar-Elena JC, Stone I (2019) Enhancing SME productivity: policy highlights on the role of managerial skills, workforce skills and business linkages. OECD SME and Entrepreneurship Papers, Issue 16.

Panizzolo R, Garengo P, Sharma MK, Gore A (2012) Lean manufacturing in developing countries: Evidence from Indian SMEs. Prod Plan Control 23(10–11):769–788. https://doi.org/10.1080/09537287.2011.642155

Sanders A, Elangeswaran C, Wulfsberg JP (2016) Industry 4.0 implies lean manufacturing: Research activities in industry 4.0 function as enablers for lean manufacturing. JIEM 9(3):811–833

Shah R, Ward PT (2007) Defining and developing measures of lean production. J Oper Manag 25(4):785–805

Sharma R (2012) Conceptual framework for improving business performance with lean manufacturing and successful human factors interventions–A case study. Int J Qual Res 6(3):259–270

Siegel R et al (2019) Integrated green lean approach and sustainability for SMEs: from literature review to a conceptual framework. J Clean Prod 240:118205

Signoretti A (2019) Explaining variation in the social performance of lean production: a comparative case study of the role played by workplace unions' framing of the system and institutions. Ind Relat J 50(2):126–149

Snee RD (2010) Lean Six Sigma–getting better all the time. Int J Lean Six Sigma

Sony M, Naik S, Antony J (2020) Lean Six Sigma and social performance: A review and synthesis of current evidence. Qual Manag J 27(1):21–36

Sternad D, Krenn M, Schmid S (2019) Business excellence for SMEs: motives, obstacles, and size-related adaptations. Total Qual Manag Bus Excell 30(1–2):151–168

Timans W, Antony J, Ahaus K, van Solingen R (2012) Implementation of Lean Six Sigma in small- and medium- sized manufacturing enterprises in the Netherlands. J Oper Res Soc 63(3):339–353. https://doi.org/10.1057/jors.2011.47

Wacker JG (2004) A theory of formal conceptual definitions: developing theory-building measurement instruments. J Oper Manag 22(6):629–650

Welsh J, White J (1981) A small business is not a little big business. Harv Bus Rev 59(4):18–32

Womack J, Jones D, Roos D (1990) The machine that changed the world: The story of lean production. HarperCollins

Womack J, Jones D (2003) Lean thinking: Banish waste and create wealth in your corporation, 2edn, ed. s.l. Simon & Schuster UK Ltd.

Yadav G, Luthra S, Huisingh D, Mangla SK, Narkhede BE, Liu Y (2020) Development of a lean manufacturing framework to enhance its adoption within manufacturing companies in developing economies. J Clean Prod 245:118726

Yusof SM, Aspinwall E (2000) A conceptual framework for TQM implementation in SMEs. TQM Mag 12(1):31–36

Zhou B (2012) Lean principles, practices, and impacts: a study on small and medium-sized enterprises (SMEs). Ann Oper Res 241:457–474

Chapter 3
Achieving Environmental Excellence Through Lean and Green in SMEs

Objectives

1. Highlight the relationship between environmental performance and Lean,
2. Guide SMEs in the implementation of a Lean and Green approach based on a real industrial case.

3.1 Introduction

The combination of the Lean approach with environmental management has gained significant momentum in recent years in the scientific community and in industrial practices (Shah and Ward 2003; Yang et al. 2010). This growing interest is quite justified, given that companies today, regardless of their size, are confronted with issues of social responsibility including the environmental aspect. About the resources consumed (water, raw material and energy) and the waste. The pressure on this subject is more felt since government policies and regulations prescribe certain recommendations and obligations. As a result, the macro and microeconomic context is pushing SMEs to consider environmental performance in their operating strategy, and to seek the methodological means to improve it.

This chapter provides some answers by asking about the combination of operational excellence management via a Lean manufacturing (LM) approach and environmental management (EM) in order to cope with the current ecological and economic situation. Moreover, scientific research through various studies (Yang et al. 2010; Dües et al. 2013; Belhadi et al. 2018b) have shown that the deployment of the Lean approach has a very positive effect on improving environmental performance. Although, this combination is widely discussed in the literature, few studies have focused on small organizations. Therefore, the content of the chapter will also feed the curiosity of SMEs on how to proceed. This will be done through the exposure of the experience of an SME that has simultaneously adopted the lean and green

F. E. Touriki et al., *Sustainable Excellence in Small and Medium Sized Enterprises*, Industrial Ecology, https://doi.org/10.1007/978-981-19-0371-7_3

approaches and that has been able to achieve very satisfactory results. This industrial case will help to clarify the minds on how to do this integration.

3.2 Lean Manufacturing and Environmental Management

3.2.1 Lean Manufacturing

The origin of lean manufacturing comes from the Total Production System (TPS). With this system, the Toyota company was able to distinguish itself by its exceptional performance in terms of quality, productivity and sales. Since then, lean has become a universal standard coveted by manufacturers.

The basic idea of lean manufacturing is to identify and distinguish between value-added and non-value-added activities. A set of methods and tools are then used to achieve the lean objective, which is the elimination of waste. Presented from this angle, the lean approach seems simple and fun.

However, the definition of lean has not obtained a consensus in the scientific community, some define it as a set of tools and methods (Ohno 1988; Drew et al. 2004), and others as a philosophy and culture (Liker 2006; Karim and Arif-Uz-Zaman 2013). In this sense Womack and Jones (2003) states that lean is more than a production system, it is a true philosophy of continuous improvement with the objectives:

a. drastically reduce waste in the supply chain
b. reduce inventory and space occupied at the point of production,
c. create more robust production systems,
d. create appropriate systems for the delivery of materials, and
e. improve the organization's production areas to increase flexibility.

In the same perspective, Shah and Ward (2007) defined lean as an integrated "socio-technical" system that continuously strives to eradicate waste by simultaneously reducing or minimizing variability from different sources. Nevertheless, the implementation of a successful lean approach requires the joint implementation of practices of the "technical system" (Just in Time, Total Quality Management, Total Productive Maintenance) and those of the "social system" (Cua et al. 2001). Indeed, as explained by (Hennion and Makhlouf 2016), the introduction of a lean transformation program within an organization, involves profound organizational and cultural changes on several dimensions: the customer, the process, the performance, the behavior and the attitude.

Furthermore, a lean production system often refers to two main pillars: just-in-time and Jidoka (Ohno 1988).

- **Just-in-time:** this is a pull production strategy, it is the customer order that triggers the manufacture of the product (Hennion and Makhlouf 2016). The latter, explain that the idea came from the systems of large supermarkets that were characterized

by a near-real time supply, adapting quickly to customer demand. So, it's about having the right products, at the right time and in the right quantity. Inspired by this operation, Toyota engineers have developed the just-in-time system in order to meet customer demand while minimizing inventory and time.
- **Jidoka:** consists of guaranteeing quality at each step of the production process, including those that are completely automated and do not require any human intervention. This involves equipping machines with systems that enable them to detect defects and stop in time. This concept is also known as "autonomation".

Another concept strongly used in lean transformation projects is Kaizen. It is nothing more or less than a synonym for continuous improvement. It is based on a series of principles (Hennion and Makhlouf 2016) that are all common sense: (1) good process leads to good results; (2) go see for yourself to understand the current situation, (3) talk with data, manage with facts, (4) implement actions to correct the root cause of problems; (5) work as a team. Also known by another name "Gemba Kaizen" to emphasize the need to be on the ground (Imai 2007). This author maintains that simple and inexpensive concrete improvements, achieved in a short time, should be implemented where the added value is created,

In addition to these concepts, there are several tools and methods that accompany Kaizen projects and that will be chosen according to the type of waste to be reduced and the objective to be reached. But, those that are considered essential prerequisites to the approach are the 5S and standardization (Imai 2007; Belhadi et al. 2018a).

Despite the spread of lean as an operational system par excellence, it remains little known and applied by SMEs despite their economic weight. Indeed Thanki and Thakkar (2018) state that over the past decades the Lean approach has in many cases allowed small and medium enterprises (SMEs) to improve their performance by providing huge opportunities for development and expansion through profitability and quality improvement. Also, many researchers have promoted the use of lean production in small and medium enterprises also Belhadi et al. (2017) and have highlighted its applicability in these companies and the possible competitive advantages obtained by lean in SMEs.

3.2.2 Environmental Management Practices

Environmental practices are diverse and varied. Companies have a wide range of alternatives for integrating environmental concerns into their strategies and daily operations. In this regard, Table 3.1 summarizes the main environmental practices according to Jabbour (2013).

It should be noted that the Environmental Management System according to the ISO 14001 standard is considered one of the most promising and effective internal management tools. Indeed, the process of certification Iso 14001 submerges the companies in an internal dynamic based on the identification of the aspects and significant impacts and their management through actions of depollution, treatment

Table 3.1 Environmental management practices according to Jabbour (2013)

Environmental Management Practice	Measures/definition
Clear policy of valorizing environmental management	Clear policy of valorization of environmental management through a precise declaration from business directors about the main environmental aspects and impacts generated
Environmental training for employees	Environmental training for all employees aimed at promoting environmental policy and permitting employee awareness of their activities' environmental impacts
3Rs (Reduction, Reuse and Recycling applied to Water, Electrical Energy and paper)	3Rs, comprising Reduction, Reuse and Recycling applied to water, electric energy, paper and other natural inputs, increasing business productivity
Development of new products with a lower impact on the environment	Development of products with smaller environmental impacts
Development of production processes with a lower impact on the environment	Development of production processes with smaller environmental impacts
Supplier selection based on environmental criteria	Vendor selection based on environmental criteria
Iso 14001 or other environmental management systems	Environmental management systems (ISO 14001 and/or others)
Voluntary promotion of environmental performance information	Voluntary promotion of information on environmental performance

and rationalization of the consumptions. This way of doing things is very relevant, because it allows to comply with the regulation, to manage the environmental risks and to increase the credibility of the company in front of its external partners. However, this proactive approach can be perceived by SMEs as costly and sometimes quite complex. Moreover, as Personne (1998) points out, the requirements of this system appear to be too high in relation to the level of environmental integration in SMEs.

Also, among the practices identified by the author, the 3Rs (Reduction, Reuse and Recycling), which seem the most appropriate and accessible in the context of SMEs. In terms of reducing the consumption of resources and reusing others. Indeed, the possibility of quantifying the resources used will facilitate the company to set reduction objectives and consequently improve its environmental performance.

In this regard, companies consider their environmental performance as a key to competitive advantage, along with cost, time and product quality (Azzone and Noci 1998). With respect to SMEs many researchers have emphasized the urgent need for SMEs to achieve environmental performance (Agan et al. 2013; Ramayah et al. 2012). However, various studies have asserted that SMEs lag far behind their larger counterparts in terms of green achievement. For example, Florida (1996) conducted a classification of companies and found that SMEs are the least environmentally efficient. Similarly Azzone and Noci (1998) based on an intensive analysis of 15 companies, found that SMEs still adopt reactive environmental strategies. Brío and Junquera (2003) came to the same conclusion and stated that SMEs are generally used to ignoring environmental issues and, at most, are content to comply with the direct regulations that affect them. In addition, Boiral et al. (2014) showed through interviews with 63 Canadian SMEs that the environmental commitment of SMEs is very limited and needs even more development.

Several authors have confirmed that the adoption of improvement initiatives such as lean within any organization, including SMEs, can directly improve environmental performance since its main purpose is to reduce waste throughout the organization (King and Lenox 2001; Dües et al. 2013). Therefore, the integration of Lean and environmental practices is recommended, especially for SMEs, as one of the key paradigms that ensure the improvement of both dimensions of a company's performance, i.e., business performance and environmental performance with less investment Thanki et al. (2016).

3.2.3 Lean and Green: Complementarity or Contradiction

According to research in this field, it has been found that common characteristics are shared between lean and environmental management, particularly in terms of waste reduction and waste. Some authors claim that the adoption of lean will inevitably lead to the reduction of pollution. Indeed, studies (Dües et al. 2013; Faulkner and Badurdeen 2014; Thanki et al. 2016; Garza-Reyes et al. 2016), have shown that the implementation of Lean can lead to significant environmental benefits.

Authors Dües et al. (2013) and Garza-Reyes et al. (2016) explained that the alignment of Lean goals with the sustainability paradigm seems natural since the goals set to achieve Lean will serve as catalysts for the successful implementation of green practices and contribute to the achievement of environmental goals. Especially since Lean hunts down different types of waste including those related to energy consumption, water consumption and through waste reduction. Seen from this angle, Lean and environmental management appear to be two complementary approaches.

However, the research work that has studied the synergies and compatibilities between Lean management and environmental performance has been carried out, once again, in the context of large companies. It is therefore difficult to bring the results for SMEs since the size of the plant is considered one of the key factors for the implementation of both Lean and environmental management practices Gandhi et al. (2018).

But these studies are very interesting in the sense that they can characterize the relationship between the two approaches. The research work has mainly raised the impact of either the Lean approach on environmental performance or the impact of the simultaneous integration of "Lean & Green" on the overall performance of the company (economic, operational and environmental). According to Yang et al. (2010) lean manufacturing is an important antecedent of environmental management practices.

According to Fercoq (2014) the two approaches have a generally positive synergy. He proposes a study on the impact of Lean and Green strategies on the various processes of the company, qualifying the synergistic character Lean and Green (see Table 3.2). This synthesis is made without specifying the size of the company or its particularities:

According to the Table 3.2, a convergence is noted in terms of means and results between Lean and Green for the activities of the realization and support processes. Also, each process can be positively impacted by the combination of the two approaches. This finding is valid for all activities except those of logistics where the author recorded a divergence of approaches at this level. However, this study excludes management processes that can have a direct or indirect impact on the achievement of operational and environmental performance.

The impact of Lean management on the company's operational performance has been widely demonstrated by the scientific community. For example, authors such as (Shah and Ward 2003) state that Lean has an impact on the performance perceived by customers (shorter lead times, lower prices, better quality). It also has an impact on financial performance by improving the efficiency of processes. In the same sense, Rahman et al. (2010) confirm that the actions of Lean (minimization of losses, acceleration of flows) are significantly correlated with operational performance (delivery time, unit cost of the product, overall productivity, overall customer satisfaction).

As for environmental management (Jabbour 2013) explain that it also improves operational performance in terms of quality, cost, lead time, flexibility, attractiveness of new products. But when used as the only approach in the company, it does not necessarily contribute to the financial performance of the company (Yang et al. 2010).

A closer look at the literature demonstrates a real synergy between Lean and environmental management and illustrates how the implementation of Lean affects environmental performance outcomes. Table 3.3 describes the recent literature, which discusses the relationship between Lean & Green.

The literature review revealed a common thread among virtually all the studies. They are based on empirical methodologies relating field experiences with the use of lean and green methods and indicators. Also, there is a convergence of the research content according to two orientations:

- **"Results"** orientation. In this case, the authors provide quantitative arguments on the impact of the prior or simultaneous use of lean on environmental performance. Using specific indicators, authors such as Vais et al. (2006), Sorbal et al. (2013), Chiarini 2014), and Pampanelli et al. (2014), assert that the environmental benefits derived from lean are mainly the effective and efficient use of resources in terms of water, raw material and energy. Similarly (Chiarini 2014) looked at the study of the environmental impacts of the processes of five large manufacturing companies before and after using Lean practices to identify the contribution and ability of these to reduce them. In addition, Dües et al. (2013), conducted a literature review on the relationship between Green and Lean. The authors came to the same conclusion: there is a positive and strong synergy between the two paradigms.

Table 3.2 Synthesis of Lean Green concepts, according to the value chain (Fercoq 2014)

	Main activities		
	Marketing, Sales and Services	**Internal and external logistics**	**Production**
Lean Management	Maximize the value added	Develop the "just in time" approach	Accelerate flows by reducing non-value added
Green Management	Green marketing (green product segmentation, promotion, eco-labeling)	Green Logistics (Consolidation of flows, Optimization of loads, sharing of own logistic means)	Resource efficiency Clean production
Convergence	**Yes**	**No**	**Yes**
Lean Green Integration	Economy of functionality-Product Service System	Reverse or closed loop supply chain/Circular economy	Production loss reduction techniques Lean Green Engineering

	Support Activities			
	Research and Development	**Purchase**	**Human resources**	**Infrastructure management**
Lean Management	Design with the customer in mind Value Analysis Concurrent Engineering	Supplier Development Plan	Development of individual and collective responsibility	Optimize the value of infrastructure processes
Green Management	Ecodesign (reduction of raw materials and energy consumption, recyclable materials, separable components, long life span)	Green purchasing (choice of materials, choice and evaluation of suppliers according to processes, logistics processes	Involvement Training Teamwork	Efficiency of utilities, waste process Integrated information systems (for Green information and for the LCA)
Convergence	**Yes**	**Yes**	**Yes**	**Yes**
Lean Green Integration	Eco-Innovation Life Cycle Assessment (LCA) Life Cycle Planning (LCP)	Lean Green Supplier Development Plan	Leader Lean Green Lean Green Engineering Feed-back using green lean KPIs	Efficiency of infrastructure processes Integrated Information Management Systems Lean Green

Table 3.3 Literature review on the Lean & Green concept

N	Reference	Research objectives	Type of firm	Findings
1	Vais et al. (2006)	Uses Lean and Green Production to improve industrial compliance with the effluent regulations and with EUs IPPC Directive under implementation in Romania, thereby improving the environmental conditions in the Bistrita River running through the town	Small, recycled paper-based mills	• The results obtained included an 87% reduction of the discharge of wastewater from 5300 to about 700m3/day, • Further results obtained were better housekeeping (5S and Kaizen) and a Total Quality Management (TQM) organization was implemented where product quality, environment and occupational health and safety are merged into one system
2	Vinodh et al. (2011)	Focuses on the exploration of various issues of sustainability using Lean initiatives	Not specified	Some of the issues that needs contemporary focus include: • Commitment to eliminating environmental waste through Lean implementation; • Identify environmental improvement opportunities; active involvement of EHS staff in planning and implementing Lean events with environmental opportunities; • Eliminate environmental waste by process improvement tools;

(continued)

Table 3.3 (continued)

N	Reference	Research objectives	Type of firm	Findings
3	Sorbal et al. (2013)	Examines how adopting Lean manufacturing practices can generate environmental benefits for a large multinational automotive manufacturer	Large multinational automotive manufacturer	• The environmental benefits gained from Lean centered on the efficient use of resources (such as water and other inputs) • In terms of enhancing resource efficiency, the most used Lean manufacturing practices at the study facility are JIT, VSM, visual management of environmental indicators, and employee training
4	Dües et al. (2013)	Explores and evaluates previous work focusing on the relationship and links between Lean and Green supply chain management practices	Not specified	• Green comes as a natural extension to Lean, • Lean manufacturers are greener than non-Lean companies, • Implementation of Green practices in turn also has a positive influence on existing Lean business practices
5	Faulkner and Badurdeen (2014)	Presents a comprehensive methodology to develop Sustainable Value Stream Mapping by identifying suitable metrics and methods to visually present them	Local manufacturer of satellite television dishes	• The approach is validated through a pilot case study conducted with a local manufacturer of television satellite dishes, • The environmental and societal metrics chosen may not be equally applicable across all industry sectors

(continued)

Table 3.3 (continued)

N	Reference	Research objectives	Type of firm	Findings
6	Pampanelli et al. (2014)	Proposes a new model, the Lean & Green Model, where the green concern for environmental sustainability is integrated with Lean thinking	Major international engineering corporation,	• Lean & Green proposed Model can reduce resource use from 30 to 50% on average • Reduces the total cost of mass and energy flows in a cell by 5–10%
7	Chiarini (2014)	Investigates whether or not Lean production tools can help reduce the environmental impacts of manufacturing companies	Large manufacturers of motorcycle component	• Measured the environmental impacts of the production processes of the case companies before and after the implementation of five Lean tools, namely VSM, 5S, cellular manufacturing, SMED and TPM, • Other than SMED, remaining four tools signaled improvements in green
8	Ng et al.(2015)	Proposes a methodology that aims to integrate Lean and Green practices, and enables the implementation of the integrated Lean and Green practices in an easy and practical manner	Large manufacturing - Metal stamped parts production	• Evaluated how Lean and green together yield synergy in improving both operational and environmental performance • Metric "Carbon-Value Efficiency" has been developed for collectively assessing Lean and Green implementation

(continued)

Table 3.3 (continued)

N	Reference	Research objectives	Type of firm	Findings
9	Piercy and Rich (2015)	Explores the broader sustainability benefits of Lean operations	Manufacturing UK-based cosmetics, pharma, metal pressing, FMCG, and furniture facilities (1 of them is small sized and the rest are large companies)	• Lean operations meet a wide range of sustainability outcomes beyond environmental benefits by developing a stage-based theoretical model of Lean sustainability, • Lean implementation and sustainability performance are interlinked
10	Verrier et al. (2016)	Gives an implementation structure to a Lean and Green methodology based on the seeking and eradication of wastes in production processes	Not specified	• The results particularly highlighted three "top" tools which, as organizational strategic drivers, may have positive effects on all Lean and Green mudas in addition to enhancing the involvement of employees, • These tools, easily accessible to less mature companies, are Genba Walk, Lean and Green VSM, Key Performance Indicators, and Visual Management

(continued)

Table 3.3 (continued)

N	Reference	Research objectives	Type of firm	Findings
11	Garza-Reyes et al. (2016)	Presents a case study where both paradigms have been combined to improve the transport operations of a world leader logistics organization in the metropolitan area of Monterrey, Mexico	World leader logistics company	• The study corroborates the positive synergies between the Lean and green paradigms, • The studied organization not only improved its operational efficiency TOVE index, and some of its individual metrics, but also its environmental performance, particularly, in relation to the reduction of gas emissions such as CO2, NOx, CO and HC
12	Gupta et al. (2017)	Presents a novel approach for assessing the wastes using a system dynamics model and validates the model in a radial tyre manufacturing case organization in India	large and reputed tyre manufacturing firm	• The model shows the overall performance of the radial tyre manufacturing unit assessed, • In addition, it throws light on the amount of greenness attainable by the organization through the implementation of Lean thinking
13	Colicchia et al. (2017)	Investigates how intermodal transport can be adopted for managing supply chains according to a Lean and Green approach	FMCG company (Procter & Gamble)	• Results confirm the opportunities to increase the sustainability and efficiency of the supply chain, in line with the extant literature that stresses the opportunities to concurrently reduce CO_2 emissions and optimize transportation costs

(continued)

Table 3.3 (continued)

N	Reference	Research objectives	Type of firm	Findings
14	Thanki and Thakkar (2018)	Investigates the impact of selected Lean and green practices on performance and evaluates the influence of Lean and green paradigms on the overall performance of SMEs	Small and Medium sized Enterprises	• Total productive maintenance is identified as the most important Lean practice, while ISO 14001 is the most significant green practice • On-time delivery and a reduction in emissions are the most critical criteria for Leanness and greenness, respectively

- "**Implementation process**" orientation. The works in this category, seek to propose lean and green implementation models to optimize benefits. Authors such as Verrier et al. (2016) have proposed a framework to link Lean and Green maturity to improve performance. This framework was developed in collaboration with several large companies to learn from their experience and thus build a knowledge base to facilitate the integration of these two approaches at the SME level. And more recently, other authors such as Ng et al. (2015), Verrier et al. (2016), Garza-Reyes et al. (2016), Gupta et al. (2017), and Colicchia et al. (2017) have proposed practical models and methodologies for integrating Lean and Green tools. All these models are validated in large companies and have shown relevant results in terms of improving environmental performance. The only model designed for SMEs is the one by Thanki et al. (2016). The model is not yet confirmed but their authors argued that it could serve as a roadmap for Lean-Green implementation, based on existing challenges for SMEs. As for the authors (Faulkner and Badurdeen 2014) made the choice to adapt a lean tool (Value Stream Map) to environmental concerns. They have therefore developed a complete methodology for building a "sustainable" value stream map by identifying appropriate metrics and methods to present them visually.

Despite the abundance of literature confirming the participation of the lean approach to reduce waste including those related to the environment (water, material and energy). Critics point out, however, that the reduction of one factor may increase another (King and Lenox 2001). According to the same author, efforts to increase the efficiency of production flows can lead to greater waste generation. Small batch production, for example, a characteristic of lean production, involves more frequent changeovers, and these changeovers may require more cleaning of production equipment and disposal of process materials.

3.3 Lean and Green Integration Models

The question of the integration of Lean and Green approaches is strongly imposed in this context. In industrial practices, particularly in large companies, several models have been developed and applied, providing very convincing results. We have chosen to briefly present three models to give an idea of this integration in a more concrete way.

3.3.1 Duarte & Cruz-Machado, 2013 Model

The authors propose a Lean and Green transformation model in the form of guidelines for the integration of the two approaches based on an in-depth study of the standards of excellence, in particular the quality awards and the Iso standards: 9001 (Quality),

14001 (Environmental Management), OHOSAS 18001 (Health and Safety) and Iso 26000 (social responsibility) (Table 3.4).

3.3.2 Pampanelli et al. 2014, Model

The Lean and Green model proposed by (Pampanelli et al. 2014) integrates "Green" practices into lean "thinking" through Kaizen projects to reduce energy and mass consumption in manufacturing cells. According to the authors, its implementation requires six prerequisites:

(1) a stable process with delivery rates above 90%, (2) a sufficient level of deployment in terms of use and application of Lean tools, (3) Employee Involvement systems in place (daily meeting, visual display…), (4) a supportive management team, (5) environmental awareness, and (6) significant use of natural resources.

The deployment model consists of five steps:

1. Value Chain Stabilization.
2. Identification of environmental aspects and impacts.
3. Measurement of environmental value streams.
4. Improvement of environmental value streams.
5. Continuous improvement.

The first step verifies the availability of the prerequisites. It is about identifying an operational cell characterized by a significant use of resources, a good deployment of Lean tools and a stable production flow.

Step (2): Aims to identify the environmental aspects and impacts related to the process in question according to the ISO 14001 standard.
Step (3): Measures the environmental value stream. In other words, it is a matter of collecting data and information on environmental indicators, particularly in terms of consumption (water, energy, consumable materials, and others) and in terms of waste and effluents.
Step (4) allows us to study the main opportunities for eliminating waste through the organization of kaizen workshops.
Step (5) concerns the development of action and communication plans in the kaizen workshops.

3.3.3 Alves 2015 Model

The model developed by Alves and Alves (2015) starts from the premise that the integration of "sustainability" considerations in the manufacturing process offers the company favorable conditions for the continuous improvement of its processes and its image in society. Therefore, they developed a model integrating lean production and sustainability principles, supported by a cultural transformation within the

Table 3.4 Guidelines for a Lean and Green Model according Duarte and Cruz-Machado, 2013

Business model criteria	Lean and green transformation
Leadership	- Management should demonstrate its commitment and involvement to a leangreen approach. They should communicate the importance of lean-green for the organization. They should establish strategic, measurable lean and green objectives - Management should ensure the principles of the lean and green approach such as waste reduction and efficiency - Management must define the commitments with stakeholders and the communications between them, as a way to reduce the environmental risk, cost and time, - Management must ensure investment to help in lean-green transformation - Management should guarantee a hierarchical structure with few levels to reduce the decision information - Management should ensure the legal requirements and govern norms - Management should ensure the application of management systems to assist in the implementation of lean-green
People	- Engaging every employee to root out lean and green waste, eliminate problems and make improvements. Defining a kaizen event as a lean-green team leader Should provide training and education in order to increase employees' skills - Apply principles such as cross-functional training, job enlargement and enrichment and flexible job responsibilities, - Organizations should encourage the employees to keep exploring new ways and suggest innovative ideas, - Organizations should ensure rewards and recognition for the employees, - Organizations should determine the necessary competences for employees
Strategic planning	- Management should ensure a clear lean-green strategy which should be shared by all levels of the organization, - Management should establish lean-green plans and objectives Strategy with focus on the customer and others considered key stakeholders, - It should be communicated through meetings and reports, - The deployment of the lean-green approach must be on a systematic way
Stakeholders	- Organizations should be focused on creating value for customers, investors, employees and communities, - Organizations should nurture a proactive and long-term relationship, - Organizations should promote commitment and communication between their stakeholders (e.g., suppliers or customers), - Organizations should define strategic alliances nurturing a close cooperation between them, - Organizations should ensure supplier selection according to lean and green criteria, - Organizations should encourage their suppliers to integrate lean-green into their business

(continued)

Table 3.4 (continued)

Business model criteria	Lean and green transformation
Processes	- Every process should add value for the customer, reducing all kind of waste (lean waste and environmental waste), - Organizations should provide the resources needed, - Management must focus on tools to enhance the success of lean-green activities - The 3R's (reduce, reuse, recycle) must be taken into consideration - Organizations should promote a better work environment applying the 6S methodology (5S + safety) - Organizations should use green value stream mapping to control better processes, taking into consideration the resources, information and waste in a current state and in a future state, - Standardization and work instruction should be defined, documenting leangreen best practices and making sure that they are followed, - Organizations should promote continuous improvement (in all the supply chain). Some tools can be used for lean-green continuous improvement such as A3 report or analytical tools
Results	- Management should select measures (lean and green measures) to see how the organization is providing value to the customer, - Organizations should monitor and measure the organization, applying lean and green metrics. These metrics can be defined as cost, productivity, quality, process, product, customers and people, - Management must analyze the data to understand the short and long term needs

company. The methodology describes specific actions to be carried out in a logical sequence to achieve sustainable results as the implementation process progresses. While developing employees' skills through continuous learning and promoting the commitment of the entire team and bring greater stability to the processes involved. The ISMA model is deployed in five steps:

1. Structuring the implementation process,
2. Planning the implementation,
3. Implementation of improvements,
4. Process Stabilization, and
5. Knowledge sharing and continuous improvement.

Phase (1) identifies the basic conditions and preparatory activities to launch the change process in a structured way. It is based on the realization of a preparatory diagnosis and the setting of organizational conditions linked to the designation of a team and a project manager.

Phase (2) focuses on the implementation of performance indicators for both the environmental and operational aspects; the realization of a value stream mapping during which the possibilities of improvement of the pilot area are identified. During this phase, the implementation of the Environmental Management System according to ISO 14001 and the Occupational Health and Safety Management System is also planned.

Phase (3) allows the implementation of the improvement programs identified in the previous phase, for example 5S, SMED, TPM…
Phase (4) concerns the stabilization of the process by consolidating the achievements. This is done through elements such as standardization of practices and the establishment of supervision and problem-solving methodologies. The last phase refers to feedback through capitalization and knowledge sharing. In other words, it is a matter of observing the results obtained and maintaining them at the desired level, refreshing the VSM, and extending the experience to other production areas.

Despite their diversity, these different models agree on a few elements:

- The need for management commitment and staff involvement,
- The initiation of the project generally begins with a diagnosis of the current state of the company by evaluating its performance via operational and environmental indicators,
- Several lean tools are deployed: 5S, VSM, Kaizen, SMED….

In relation to this last point, the literature review of the corpus of literature has enabled us to identify the lean tools and approaches most involved in supporting the improvement of environmental performance within companies. Table 3.5 summarizes these tools.

The authors cited above converge on the idea that the 5S, which is a Japanese method based on simple but strict rules in terms of tidying, cleaning and visual management, has a direct impact on improving environmental performance. Indeed, this method ensures a safer workspace, and encourages employees to follow standardized procedures and maintain a clean workplace which reinforces the proper disposal of waste (Sobral et al. 2013). These same authors state that the most used Lean practices in terms of improving resource efficiency are JIT, VSM, visual management of environmental indicators and employee training. This is supported by several authors such as Dües et al. (2013) and Faulkner and Badurdeen (2014) who stated that Lean practices serve as a catalyst to facilitate the implementation of environmental management. For example, they pointed out that CO_2 emissions can easily be added as an additional source of waste in the VSM approach. They added that JIT can help achieve gains in the areas of inventory reduction and small batch production, which could lead to greater reduction in material and energy waste. For example, Sobral, et al. (2013) explained that after implementing JIT, one company recognized that they could now receive parts from their supplier without a protective layer of oil, since the parts would be stored for a short period of time. This eliminated the need to wash the parts on the production line, thereby reducing water consumption.

Other Lean practices are also discussed: Vais et al. (2006), Vinodh et al. (2011), and Piercy and Rich (2015) noted that TPM encourages preventive and proactive maintenance of equipment to improve its life. This increases equipment longevity and prevents process failures that cause leakage, scrap, rework, and wasted raw materials. In addition, SMED and Kanban are considered strategic organizational factors, which have a potentially beneficial influence on all Lean and Green mudas (Verrier et al. 2016). In addition, Chiarini (2014) measured the impact of several

Lean factors on environmental performance and noted that manufacturing according to the U-cell organization could result in lower electricity consumption.

3.4 Can SMEs Benefit from This Synergy?

From the above, it is raised that the synergy between lean manufacturing and environmental management, is undoubtedly demonstrated by several studies. Indeed, Lean and Green approaches make a positive contribution to sustainable excellence via the improvement of economic, social and environmental performance. In this sense, Gandhi et al. (2018) have confirmed that trade-offs between Lean and environmental management paradigms could help manufacturing SMEs to become more efficient and sustainable. On the other hand, evidence suggests that organizations (including SMEs) have encountered difficulties in integrating and implementing these two approaches. Nevertheless, little research has provided a proven and practical way to integrate the two approaches, combining their foundations and principles, in the specific context of SMEs. Given this gap in the current literature, questions arise: "How can Lean practices be used as a catalyst to achieve environmental performance? In other words, how can environmentally friendly manufacturing be developed to generate profits in the context of SMEs?

Because of this, and to accompany companies especially SMEs, a real case of an industrial SME manufacturing pumps will be developed based on the work of Belhadi et al. (2018b). The interest of the case study lies in the fact that it presents a methodological framework proposing a structuring according to three phases and thirteen steps, guiding manufacturers in the integration and implementation of Lean and Green approaches.

3.4.1 Presentation of an Industrial Case

The company, in question, is an SME created in 2014, employing 12 Executives, Engineers, and 154 employees. Its main activity is the production of a wide range of industrial pumps such as centrifugal pumps, multistage submersible pumps, clamshells, drain valves meeting all the production needs of phosphate and phosphoric acid manufacturing industries.

Its current production capacity is 20,000 centrifugal pumps per year and 1,500 deep well pumps per year, in addition to submersible pumps, high pressure industrial pumps and domestic pumps of different designs and capacities. It also manufactures sluice gates and non-return valves (Reflex) with diameters ranging from 37 to 200 mm.

It is worth noting that the company has obtained ISO 9001 version 2015 certification for its quality management system, as the first and only pump and casting industry in Morocco. Although the company is part of a larger industrial group, it acts as an independent small and medium-sized company in North Africa, responsible for its own activities and operations.

The company has several cells, each responsible for a basic operation. The pump manufacturing process consists of four stages: Casting, Machining, Assembly and Testing for which there are several work procedures to produce a ready pump (Fig. 3.1).

Foundry Cell: produces metal castings. Metals are cast into shapes by melting them in a liquid, pouring the metal into a mold, and removing the molding material or casting after the metal has solidified as it cools. The most processed metals are aluminum and cast iron. However, other metals, such as bronze, brass, steel, magnesium and zinc, are also used to produce castings in foundries. In this process, parts of desired shapes and sizes can be formed.
Machining cell: is the process to make the desired shape of the material by several machining operations. The company has different machines to do this task such as lathe machine, milling machine, forming machine, grinding machine, and drilling machine.
Assembly and painting cell: is the last part of the manufacturing process. It includes the installation of mechanical components (bearing, shaft, mechanical seal, flange coupling) and hydraulic components (impeller, volute), then the installation and tightening of bolts. Sometimes maintenance work is also performed in the assembly cell.
Test and quality control cell: After assembly, a pump goes through the test and quality control cell to confirm its readiness for shipment to the end customer. If not, the pump is sent back to the assembly cell to make the necessary adjustments.

Freshly integrated into the highly competitive pump market, the company has realized that it must provide high quality products while reducing production costs and delivery times in order to remain competitive. On the other hand, the company faces strong pressure to improve its environmental performance from its shareholders and business partners. In addition, most of its customers are currently large companies that are starting to apply criteria for selecting their suppliers. By indicating to them the importance of respecting environmental performance standards if they want to continue sourcing from them.

Apart from its external pressures, the company's internal situation suffers from a number of serious problems in its manufacturing plant:

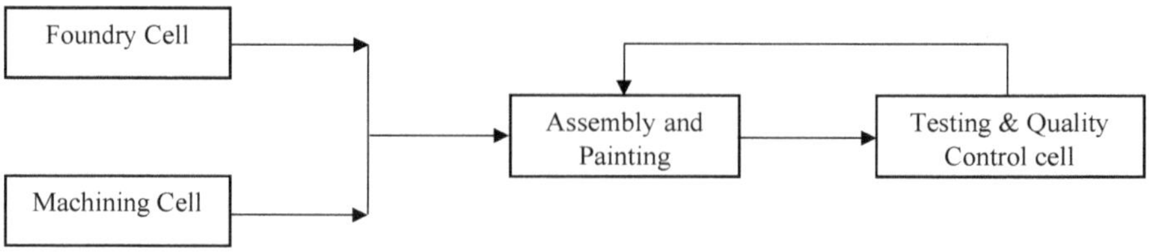

Fig. 3.1 Flowchart of the process of pumps manufacturing

- The cost of starting up new facilities is very high,
- Additional production costs are estimated at 20%,
- Excessive consumption of raw materials and waste: the company consumes 40% more raw metals and 65% more water than its competitors,
- Cycle times are often three times longer than expected,
- Energy consumption is doubled in three years,
- Customer complaints increase drastically by more than 10% each year,
- Emergency repairs and unscheduled downtime exceeded 20% of overall uptime.

Operating under these conditions, the company found itself strongly challenged to improve its performance in different ways and to meet the demands of its stakeholders. As a result, it decided to adopt a new management approach by engaging in a Lean and Green transformation to improve both its operational and environmental performance relative to its competitors.

3.4.2 Implementation of the Lean and Green Approach

The integration model used is that of Belhadi et al. (2016) which is adapted to the context of SMEs. It has been followed in its entirety for this case study, obviously, the choice of lean tools has been adjusted to consider the environmental dimension. Figure 3.2 presents the detailed methodology in 3 phases.

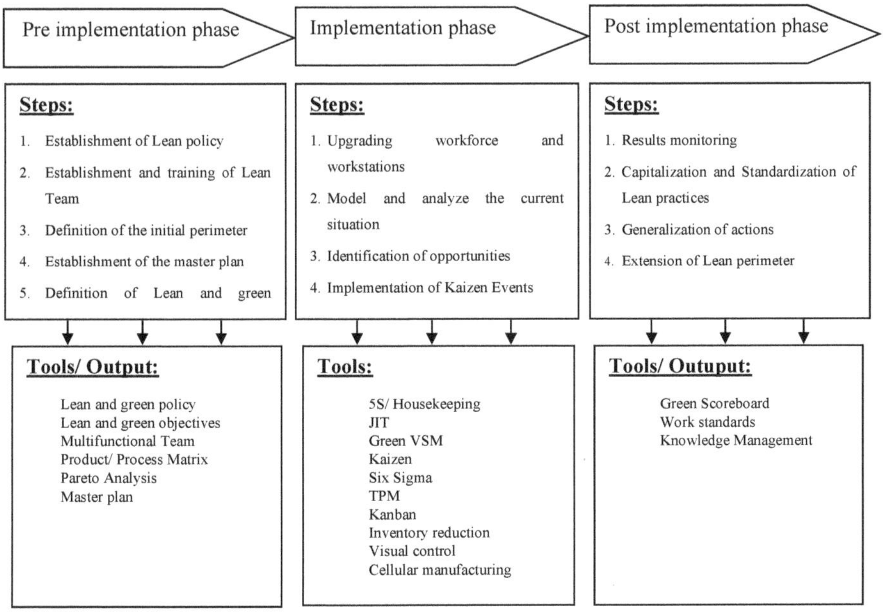

Fig. 3.2 Lean implementation phases adjusted to Environmental Management

The phases were carried out as follows.

3.4.2.1 Pre-Implementation Phase

Step 1-1: Establishment of a policy with Lean objectives and environmental performance

It is important to remember that the involvement and commitment of the management in any improvement project is fundamental if it wants the participation and cooperation of its employees. This commitment can be expressed through the drafting of the policy and the definition of objectives. In our case, the management has established a policy according to its global strategy, including the following objectives:

1. Increase production capacity and quality,
2. Reduce production costs by optimizing the use of materials and energy.

The management together with the operators, engineers and managers of different departments conducted meetings to ensure wide dissemination of the policy.

Step 1-2: Designation and training of the Lean team

A "Lean team" is designated within the company and composed of a process engineer, process technician and skilled operators. The structure is illustrated in Fig. 3.3.

The company director is considered the sponsor of this team, i.e., he is responsible for supporting the project and providing the necessary technical, human and financial resources. The process engineer, named "Lean Leader", must supervise the creation of flow maps, ensure the training of his team on the Lean concept, its philosophy and its tools, and accompany the implementation of Kaizen projects. In addition to the other members (process technician and skilled operator), the authors of the case study are integrated into the Lean team as consultants.

After having a functional team, this one benefited from a training, not only to understand the lean concept and to be able to apply these tools, but the goal was also, to acquire a lean culture and to be able to spread it to the other collaborators. This

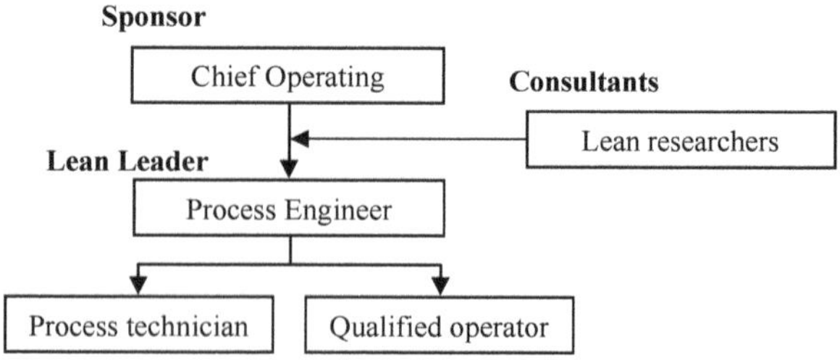

Fig. 3.3 Structure of the Lean team

was done in order to accompany the employees during the transformation and thus reduce the resistance to change while creating a favorable work environment.

Usually, in the case of large companies, training is provided by consulting firms with external professional trainers. However, since this is an SME and given its limited resources, the training is conducted by the Lean leader and his consultants (case study writers).

Step 1-3: Define the initial scope

The opportunities for improvement are virtually unlimited, but they must be managed with very limited resources given the size and nature of the company. Therefore, the scope of the project is crucial.

The scope of the Lean project is determined by identifying the priority value stream. The team can then demonstrate the effectiveness of the approach through the achievement of convincing results, and this is even a source of motivation and pride for the personnel. Then, once the experience is acquired, its duplication can be facilitated because the employees now know the ins and outs of the Lean and environmental management approaches. The scope will then serve as guidelines for future projects. The criteria for the selection of the scope were developed as follows:

- The largest value stream in terms of production volume and annual turnover,
- The simplest value stream with minimal equipment.

Finally, a Pareto analysis was performed to identify which value stream to target first.

Step 1-4: Establish the Master Plan

Table 3.6 summarizes the estimated time in days for each of the Lean steps. The total estimated duration of the project is 156 days, not including the last two stages of the process: generalization and extension of the Lean perimeter.

Step 1-5: Define Lean and Environmental Management Indicators

Measurement is an essential component of continuous improvement systems. It is necessary to establish appropriate measures and indicators to visualize over time the improvements achieved. Therefore, and in accordance with the objectives stated in the policy, a number of indicators related to Lean and environmental management are defined. Table 3.7 shows the proposed parameters.

The indicators concerning Lean in its operational dimension, particularly consider the sources of waste. Thus, four measures are considered: efficiency, quality rate, stock rate and availability rate.

As for the environmental issue, the input and output elements of the manufacturing process are taken into consideration to reduce the environmental impacts of the

Table 3.5 Lean tools with impacts on environmental performance

Lean practices	Impact on Green Performance	1	2	3	4	5	6	7	8	9	10	11	12
5S	Ensure that employees follow standardized procedures and maintain a clean workplace, which strengthen the proper disposal of rejects and incorrect use of inputs	x	X	x			x	x	x	x	x	x	
JIT	Ensure receiving parts from supplier and production when needed. This eliminates the excess consumption of energy for storage			x	x								
TPM	Encourage preventive and proactive maintenance of equipment to enhance its cycle life, increased longevity of equipment and prevent process failures that cause leaks, scrap, rework and raw materials wastage	x	X	x			x			x	x	x	
VSM	Showcase the potential for addressing wastes through fewer defects, less scraps, low energy usage		X		x	x	x		x	x	x	x	
SMED	Promotes small lot sizes through quick change over time which reduces the energy consumed during storage and handling of big lots of products						x		x	x	x	x	
Kanban	Reduces WIP and inventory thus reduces the energy used for storage and transport of inventories								x	x	x	x	
Visual control	Systematizes the identification and elimination of unwanted entities hence less material usage and wastes		X			x				x		x	
Cellular manufacturing	Promotes the reduction in set-up times and change over time thus achieve low energy and resource consumption along with low level of defects		X	x			x						

(continued)

Table 3.5 (continued)

Lean practices	Impact on Green Performance	1	2	3	4	5	6	7	8	9	10	11	12
Inventory reduction	Facilitates identification of failures in the organization's processes, thus helping it avoid generating excess consumption and waste			x	x				x				x
Kaizen	Strengthens and handles the development and sustainability of an environmental management system aiming at eliminating hidden wastes and unwanted activities	x	X	x				x	x				
Six Sigma	Ensures lower level of defects thus less rework and waste, improvement in product durability and reliability hence enhance product life cycle	x	X										

(1) Vais et al. (2006), (2) Vinod et al. (2011), (3) Sorbal et al. (2013), (4) Dües et al. (2013), (5) Faulkner and Badurdeen (2014), (6) Chiarini (2014), (7) Pampanelli et al. (2014), (8) Ng et al. (2015), (9) Piercy and Rich (2015), (10) Thanki et al. (2016), (11) Verrier et al. (2016), (12)Gupta et al. (2017)

Table 3.6 The substages and deadlines estimated for their completion

Phases	Steps	Deadlines
Pre implementation phase	• Establishment of Lean Policy/ Lean objectives	1 Day
	• Establishment and training of Lean Team	7 Days
	• Definition of the initial perimeter	2 Days
	• Establishment of master plan of Lean deployment	2 Days
	• Definition of Lean and green indicators	5 Days
Implementation phase	• Upgrading workforce and workstations	30 Days
	• Modeling and analyzing the current situation	7 Days
	• Identification of opportunities	2 Days
	• Implementation of Kaizen Events	90 Days
Post implementation phase	• Results monitoring	5 Days
	• Capitalization and Standardization of Lean practices	5 Days
	• Generalization of actions	In continuous
	• Extension of Lean perimeter	In continuous

company. But, in our case, only the input elements will be measured and followed and which concern: the energy consumption, the use of raw materials and the water consumption. The output elements in terms of pollution and waste will not be taken into account due to the lack of resources to limit this type of impact. However, it is always a good idea to measure these kinds of indicators at least to become aware of the degree of environmental impact, even if improvement paths cannot be taken due to the lack of resources and technical expertise in this area. The measurement of consumption will identify the number of resources consumed per process to produce a part as well as the energy consumed between the transport and/or storage processes.

3.4.2.2 Implementation Phase

Step 2-1: Upgrading Workforce and Workstation

The momentum of the Lean transformation will only effectively begin with an upgrade program to create an environment conducive to culture change (Belhadi et al. 2016). This program targets two fundamental resources:

- **Human resources:** a training plan is conducted on Lean principles and practices for all employees. In addition to a communication and awareness campaign explaining their role and contribution in achieving environmental performance.
- **Material resources:** a maintenance and cleaning program is initiated to reorganize the workplace in the machining area because the degree of disorganization was too high in this space. Therefore, it was decided to start with this area by

making it a pilot area. As a result, manufacturing in the machining cell was interrupted for an entire day. This sends a strong message to all employees about the management's commitment to this project. All the workers in the machining area, about 17 people, participated in the event.

The 5S method composed of 5 steps is deployed in the machining cell began by:

The "Sort" step: Each operator sorted out the useful and useless objects considered not necessary for the execution of their tasks in their workstation.
The "Straighten" step: Visual labeling is used for work tools with the help of color coding to distinguish useful and useless items in the production area. Then, each item and tool is put in the appropriate place near the workstation using it. This is done to reduce the time wasted in moving and searching for the tools needed for the task.
The "Shine" stage: once each item is in its place, this stage establishes a discipline related to the cleanliness of the premises by cleaning the floor, equipment and tools. Subsequently, work standards and best practices are formalized in procedures that are readily available to all employees.
The "Standardize" step: allowed us to clearly define the locations of each necessary element using visual indicators. Such as lines and signs that are used to identify crosswalks and different zones.
The "Sustain" step: This is the last step in the 5S method, and its purpose is to sustain the achievements of the previous steps. To this end, a "GEMBA" field visit by the management is planned every week. This allowed, on the one hand, the managers to make sure that the 5S are permanently maintained in the area and that the procedures and notice boards are kept up to date. On the other hand, it was an excellent opportunity to encourage the operators to make suggestions for improving the process and the environment of their workstation. In fact, several improvements were implemented, such as the identification and sorting of waste, and the installation of a new automatic lubrication system for the machining equipment. This has resulted in a significant reduction in leaks between the equipment and the floor. As a result, after one month of implementing the 5S method, it was found that no leaks are noticeable in the machining area.

These housekeeping programs were really the most insightful way to make employees aware of the environmental impacts and sources of raw material waste. Subsequently, a form of competition developed between operators to have the least amount of waste in their workstation.

Step 2-2: Modelling and analysis of the current situation

After having reached an acceptable level of culture in terms of Lean and particularly visual management within the company, the mapping of value streams could be started. Thus, an extended and adapted version of VSM was used to integrate environmental indicators with operational ones.

The first meeting of the Lean team on the shop floor took place with very simple means (A3 paper, pencil, eraser) to draw the VSM of the current situation. The process technician and the skilled operator started by observing the attitudes of the operators. They ensured that all operations were performed according to standard work instructions. Next, the team followed the physical flow and took the necessary times for each operation. The involvement of the operators at this stage is essential to the success of this work. For this reason, the team leader thoroughly explained to each operator the activity in progress to ensure their commitment. Tables 3.8 and 3.9 illustrate the results of the process information on the operational and environmental aspect.

From these data, it can be seen that the Foundry cell has the highest stock and changeover rates (4.6% and 32 min respectively), the Machining cell has the lowest availability rate (81.77%) and the Assembly cell has the highest quality defect rate (9.38%).

From an environmental point of view, a high total consumption of water (calculated from the daily production of 29 pumps) can be observed, especially in the foundry and painting cells. In terms of energy consumption, it can be observed that the transport of products and materials between cells consumes about 9.8% (2120

Table 3.7 Lean and Green Mertics

	Metrics	Unit	Inputs	Calculation
Productivity and quality metrics	Rate of Time efficiency	%	• Value Added Time • Cycle Time	$= \frac{\text{Value Added Time}}{\text{Cycle Time}}$
	Rate of Quality defect	%	• Total of defects • Total of products	$= \frac{\text{Total of defects}}{\text{Total of products}}$
	Rate of inventory	%	• Total inventory • Total sales	$= \frac{\text{Total inventory}}{\text{Total sales}}$
	Rate of Availability	%	• Up time • Total Time	$= \frac{\text{Up Time}}{\text{Total Time}}$
Resources and energy consumption metrics	Specific consumption of water	m^3/product	• Volume of water consumed • Total of products	$= \frac{\text{Volume of water consumed}}{\text{Total of products}}$
	Specific consumption of energy	Wh/product	• Amount of energy consumed • Total of products	$= \frac{\text{Amount of energy consumed}}{\text{Total of products}}$
	Specific consumption of crude metals	Kg/product	• Amount of crude metals consumed • Total of products	$= \frac{\text{Amount of crude metals consumed}}{\text{Total of products}}$

Table 3.8 Information needed for operational indicators

	Foundry	Machining	Assembly	Painting	Testing & QC	Total
N° of operators	1	1	1	1	1	5
Cycle Time (C/T) (min)	21.44	19.05	24.22	16.15	7.71	88.57
Total Time / day (min)	960	960	960	960	960	960
Changeover Time (C/O) (min)	32	9	10	8	2	61
N° of products/ day	41	56	32	78	70	277
N° of good products/ day	39	54	29	73	70	265
Total inventories	178	106	91	211	189	775
Up Time (U/T) (min)	879	785	935	948	960	901.4
N° of defects	2	2	3	5	0	12
Value added Time (min)	9.01	9.11	14.02	6.80	4.46	43.4
Rate of Time efficiency (%)	42%	48%	58%	42%	58%	**49%**
Rate of Quality defect (%)	4.88%	3.57%	9.38%	6.41%	0.00%	**4.33%**
Rate of Inventory (days)	4.6	2.0	3.1	2.9	2.7	**15.3**
Rate of availability (%)	91.56%	81.77%	97.40%	98.75%	100.00%	**93.90%**

Table 3.9 Information on the environmental indicators of the process

	T F Cell	**F Cell**	**T U Cell**	**U Cell**	**T A Cell**	**A Cell**	**T P Cell**	**P Cell**	**T T&C Q Cell**	**T &C Q Cell**	**T S**	**Stock**	**Total**	**Specific consumption**
Water consumed (m3)		**5**		0.16		0.07		**12**		0.05		0.05	17.33	**0,6 m3/ product**
Energy consumed (Wh)	**350**	6000	**120**	5300	**500**	1800	**150**	3500	**500**	2890	**500**		21610	**745 Wh/ product**
Crude metals consumed (Kg)		810		900		180							1890	**65.17 Kg/ Product**

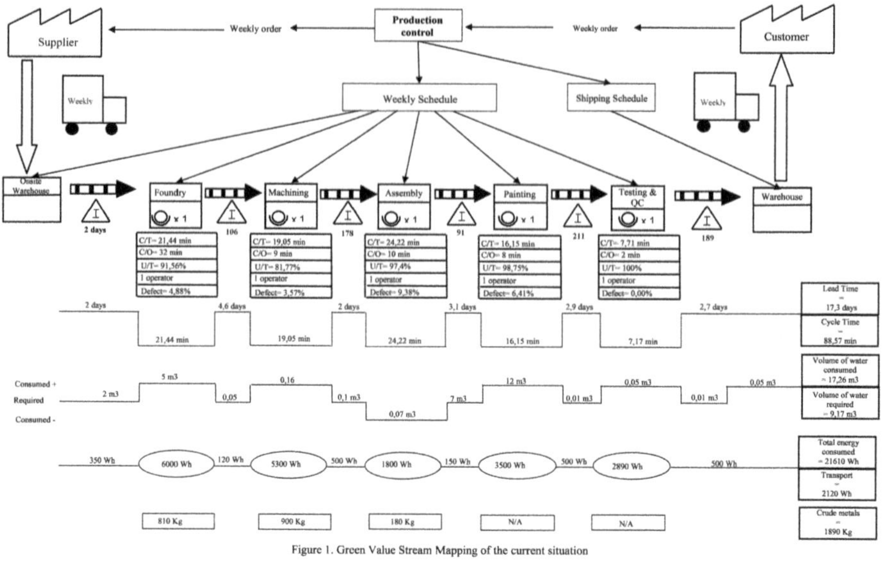

Fig. 3.4 Green Value Stream Mapping of the current situation

Wh) of the total amount of energy used (21,610 Wh). In addition, the consumption of the metal raw is absolutely high considering that the total mass of metal in a product is 40 kg while the consumption of metal raw is about 65.17 kg/product.

Figure 3.4 shows (Green VSM) the mapping of the value chain with the aspect of environmental management.

Step 2-3: Identify opportunities for improvement

The Green VSM of the current process situation has clearly shown that the manufacturing line suffers from a significant number of sources of waste. They are particularly related to inventories, overproduction, unnecessary transports and quality defects. Regarding the environmental indicators, the excessive use of materials, energy and water is noted.

To rectify the situation, the objective of this step is to redesign the future VSM through the identification of improvement path. As a result, the team is engaged in various improvement paths:

1. Synchronization of the manufacturing flow to Takt Time,
2. Implementation of a continuous flow,
3. Implementation of a pull flow at the points where the continuous flow is interrupted,
4. Identification of the pacemaker,
5. Calculation of the management delay.

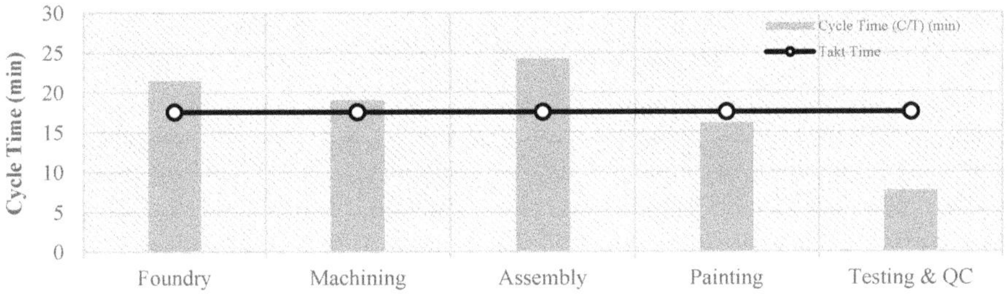

Fig. 3.5 Current situation of the cycle times in comparison with the Takt time

1st Improvement Action: Synchronization of the Production Flow with the Takt Time

Takt time is defined by Rother and Shook (1998) as the time that the company should produce a part of the product according to the sales rate in order to meet the customers' demand. In our case, and with an annual demand of 20,000 pumps and an overall working time of 16 h per day, the Takt Time can be calculated as follows:

$$\text{Takt Time} = \frac{\text{Annual working Time}}{\text{Annual demand}} = \frac{16 * 365 * 60}{20000} = 17.52\,\text{min} \tag{3.1}$$

After a graphical representation of the results obtained (Fig. 3.5). it is noted that the cycle times of the foundry, machining and assembly cells exceed this Takt time. To adjust them, a set of Kaizen events have been proposed:

1. Reduction of the changeover time in the foundry station,
2. Reduction of the downtime in the machining station and
3. Quality improvement in the assembly station.

2nd Improvement Action: Implementation of a Continuous Flow

Continuous flow refers to the production of one part at a time, with each item moving immediately from one process step to the next without stagnation in between. In order to establish a continuous flow, the team observed that the foundry, machining, assembly, and painting stations have relatively similar cycle times, which is approximately Takt time. Thus, a continuous flow can be established between these sections. A Kaizen project of cellular manufacturing incorporating these stations is proposed (see next step).

3rd Improvement Action: Establish a Pull Flow in Places Where the Continuous Flow Is Interrupted

In terms of time, the Testing & QC stations are faster than the other stations. Therefore, continuous flow may not be possible. According to the recommendations of Rother and Shook (1998), it will be relevant to link it with the proposed cellular manufacturing between the other stations through a supermarket-based pull-flow system. For this purpose, it is proposed to use Kanban cards. The goal is for the test and control station to go to the supermarket and pull the required quantity and rate of production. In turn, cellular manufacturing makes what is pulled. This allows for production control between cellular manufacturing and the test and control station.

4th Improvement Action: Identification of the Pacemaker Process

In order to control production throughout the chain and set the pace of the production flow, it is necessary to select a point in our door-to-door value chain called the "pace maker" process (Rother and Shook 1998). For this purpose, testing and quality control are selected as the pacemaker process for several reasons: First, there is no supermarket in the downstream process. Second, it is the most downstream process in our door-to-door value chain.

5th Improvement Action Calculation of the Management Time

Management time or "Pitch" is the time that determines how often production is monitored and scheduled to respond immediately to anomalies (Rother and Shook 1998). The calculation of this time is as follows: 17.52 min Takt time $\times$ 4 pumps (Since centrifugal pump customers buy in multiples of four pumps) therefore, Kanban $=$ 70.08 min. This implies that every 70 min the production is scheduled, checked and adjusted to its rhythm.

Based on the information gathered from the analysis presented above, the required improvement opportunities are marked on the future status of the Green VSM shown in Fig. 3.6 as Kaizen events.

Step 2-4: Implementing Kaizen Events

Achieving the future state requires a commitment at all levels to implement the proposed set of Kaizen events. To this end, all the necessary resources are made available to the lean team responsible for accomplishing the action plan presented in Table 3.10. It should be noted that these Kaizen events were implemented simultaneously by different Kaizen teams.

The first Kaizen event is designed to create "cellular manufacturing" between the foundry, machining, assembly and painting cells to promote continuous flow and reduce WIP. Based on space constraints and interconnections with other operations, the Kaizen team developed the cellular design shown in Fig. 3.7. Then, the construc-

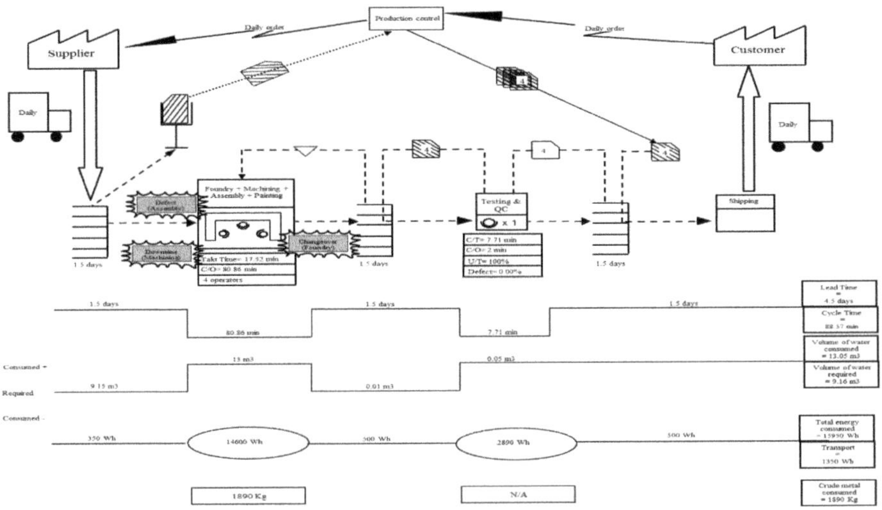

Fig. 3.6 Green Value Stream Mapping de la situation future

Table 3.10 Action plan of future situation accomplishment

N°	Station	Value Stream Objective	Goal	Kaizen Event	Timeline
1	Foundry + Machining + Assembly + Painting	Continuous flow Low inventories	0 WIP <= 1.5 days	• Cellular manufacturing • Supermarket pull system and Kanban	22 days
2	Foundry	Eliminate C/O	<= 10 min	SMED methodology	15 days
3	Machining	Increase U/T	>90%	Autonomous maintenance (first 3 steps)	30 days
4	Assembly	Eliminate defects	<5%	Quality and Process control	22 days

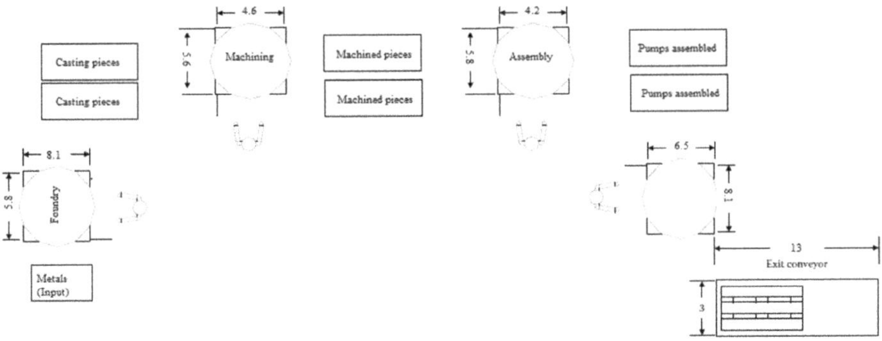

Fig. 3.7 Design of the cellular manufacturing implemented

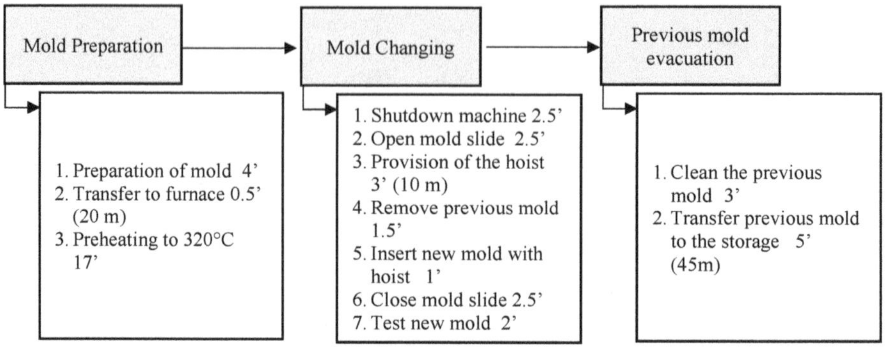

Fig. 3.8 Foundry setup operations with information regarding durations and distances

tion of the cell began with cleaning and painting, installation of equipment and tools, and some maintenance tasks. The use of rotating lifting tables was necessary to improve the flexibility. The four operators of this new installation are trained in all the operations of the different units, which leads to the following advantages:

a. The operators become more flexible and versatile, which breaks the monotony of the work,
b. Waiting time and WIP are eliminated. This has generated a huge reduction in inventory, space used, energy and resources consumed.

The second Kaizen project is intended to implement the SMED method in the Foundry unit. The SMED is a method that consists in analysing the manufacturing processes and then reducing the materials, resources and time required for the implementation (Ortiz 2006). This action is justified first, from the observation noted in Table 3.8 that the Foundry cell detects the time of change of series the highest (32 min) compared to other stations. Also, by the fact of installation of the continuous flow which does not tolerate times of high changes. To do this, the Kaizen team used the famous ESEE model for equipment changeover. This is an acronym for a sequence of Eliminate, Separate, Externalize, and Execute activities (Henry 2013).

a. Eliminate all unnecessary or non-value-added tasks,
b. Separate external and internal tasks,
c. Outsource all tasks that can be done during production,
d. Execute by performing all changeover tasks repeatedly with minimal variation.

The implementation of SMED started with the understanding of the changeover operations in the foundry station through the identification of the tasks and tools used. This activity was carried out using information from the operators in the field on the one hand, and video recordings allowing direct observation of preparation, assembly and adjustment operations on the other. The results of this analysis are summarized in Fig. 3.8.

After understanding the changeover operations, it is essential to separate the internal and external operations. To do this, checklists have been used to include all

Table 3.11 The study and the conversion of internal operations to external operations

Operations	Duration (s)	Current Nature	Future Nature
Preparation of Mold	240	External	External
Transfer to furnace	30	External	External
Preheating to 320 °C	1020	Internal	External
Shutdown the furnace	150	Internal	Internal
Open mold slide	150	Internal	Internal
Provision of the hoist	180	Internal	External
Remove previous mold	90	Internal	Internal
Insert new mold with hoist	60	Internal	Internal
Close mold slide	150	Internal	Internal
Test new mold	120	Internal	Internal
Clean the previous mold	180	External	External
Transfer previous mold to the storage	60	External	External
Total of Operations	2430		
Total of the internal operation (Current state)	1920		
Total of the internal operation (Future state)	620		

necessary information about the changeover tasks and operations. Before proceeding to the next step, an evaluation of the transport method must be performed. The objective is to reduce the time required to move parts, components and accessories.

The third step of SMED is summarized in Table 3.11 summarizes the study and conversion of internal operations into external operations. It should be noted that the changeover time should consider only the time of internal operations, because external operations are performed during the execution of the foundry activities. As a result, it can be widely observed that the conversion of the initially internal 320 °C preheating operation will significantly reduce the changeover time from 1920s (32 min) to 620s (10.33 min).

In the last step, the team brainstormed solutions to further reduce the changeover time through a planning action. It consisted of executing the opening of the mold slide and stopping the oven in parallel. This reduced the changeover time by 150 s. Figure 3.9 shows the final sequence of changeover operations.

Only with good planning and preparation, and without major investment, can the changeover time be reduced from 1920s (32 min) to 570 s (9.5 min).

Third Kaizen project related to the implementation of an autonomous maintenance program in order to reduce the machine downtime in the machining cell with the lowest availability rate. This action is always linked to the fact of the installation of the continuous flow which requires a very high rate of availability of the equipment. To do this, the implementation of 5S beforehand played an important role. Indeed, during its execution, the machining operator had the opportunity to proceed to a thorough inspection (including inaccessible places) of the machine to detect anomalies (sources of contamination and quality defects). Then, an action plan is

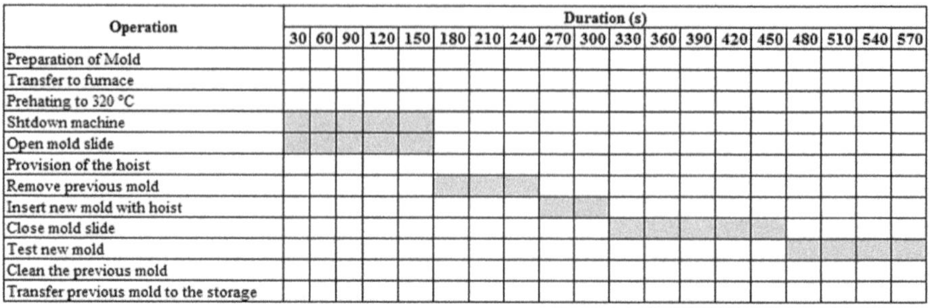

Fig. 3.9 Planning of the final sequence of the changeover operations

Table 3.12 Countermeasures of the establishment of the basic

Categories of abnormalities	Root causes identified	Countermeasures
Loose bolts on the cover	• Intensive vibration • Misalignment of drive system	• Measure of vibration • Alignment of drive system • Bearing changing
Leak of coolant water	• High pressure in coolant line	• Safe unit installation to control pressure
Heating of electric terminal box	• Overvoltage • Malfunction of electrical safety materials	• Installation of transformer to adapt electrical power • Rehabilitation of electrical safety materials
Oil leak	• A worn coupling in the oil pipe • Deterioration of the pipe	• Changing of the worn coupling • Changing of the pipe
Presence of metal debris	• Lack of lubrication	• Sealing of the oil pipes

established by a machine specialist to deal with the anomalies and contaminants found inside (dust, stains or grease). Finally, the equipment was cleaned to restore its baseline condition. Table 3.12 provides an overview of some of the countermeasures implemented.

After establishing the basic conditions of the equipment, cleaning, oiling and checking standards were developed and combined with the previously established 5S standards.

Fourth Kaizen Project concerns the reduction of the high rate of defects in the assembly cell which hinders the continuous flow by scrap and rework. Some members of the Lean team were trained in defect analysis. Then, a data collection via the 3 months history took place to determine the most important defects. The results are presented in the Fig. 3.10.

Faced with the two major defects "bearing defects" and "poor lubrication of seals" representing 70% of the total defects, an analysis of the root causes was conducted followed by a proposed action plan to eradicate them. Table 3.13 summarizes the causes identified and the improvements made.

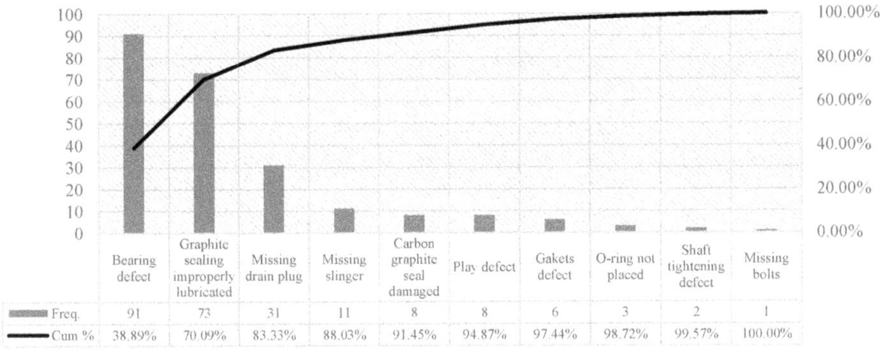

Fig. 3.10 Pareto chart of quality defects - Assembly unit

The occurrence of a high number of bearing defects, is explained on the one hand by the use of an inappropriate type and quality and on the other hand by the inadequate cleaning condition of the storage area and the workplace. For this reason, the team proposed the use of another type of bearing to reduce rotational friction and support radial and axial loads, which is the ball bearing. In addition, the suppliers of this type of bearing were diversified in order to compete and ensure the best quality. On the other hand, the design and cleaning of the bearing storage area and the workstation have prevented damage caused by particles entering the bearing. Regarding the poor lubrication of the graphite seal, it was observed that the use of silicone instead of glycerine or light machine oil is the main cause of this defect. In addition, contamination of the lubricant by foreign matter can also cause this defect. Therefore, the use of an appropriate lubricant and proper packaging of the lubricant by covering the storage tanks were found to be effective in solving the problem.

3.4.2.3 Post-Implementation Phase

Step 3-1: Tracking Performance Indicators

After the implementation of all the necessary improvement plans, the Lean team monitored the results by recalculating the previously defined indicators to show management and all employees involved the usefulness of the project. As previously stated in the Lean and Green policy, improvement must be monitored both operationally and environmentally. Table 3.14 summarizes the comparison of Lean and Green indicators before and after the implementation of the improvements.

On the operational level: It is clearly demonstrated that all productivity and quality indicators have improved considerably. These improvements have been achieved, without major investments, through targeted actions with strong staff involvement and participation:

Table 3.13 Root causes and improvements to the most significant defects in the assembly station

Defects	Root causes	Improvements
Bearing defect	Inappropriate use of Roller bearing to support radial and axial loads	• Use of Ball bearing
	Poor quality of bearing used	• Diversification of bearing suppliers • Test of different brands of bearing
	Inappropriate storage of bearing	• Arrangement of bearing storage zone in warehouse
	Presence of foreign bodies, moisture and contamination in the assembly place	• Cleaning of bearing, shaft and workstation during bearing insertion • Standards of cleaning
Graphite sealing improperly lubricated	Use of silicon lubricants	• Replacement of the current lubricant by Water, glycerin or lightweight machine oil
	presence of foreign objects in lubricant	• Covering tanks of lubricant storage • Cleaning the workstation while applying the lubricant

- First, the time efficiency rate increased from 49% to about 62% thanks mainly to the reduction of waiting time via the creation of the continuous flow between the foundry, machining, assembly and painting cells. In addition, the changeover time was reduced from 32 min to 9.5 min in the foundry cell.
- Second, the quality defect rate decreased from 4.33% to 3.61% after the implementation of the improvements. This decrease is due to two reasons: the significant decrease of the defect rate in the assembly cell from 6.38% to 3.13% and the increase of the daily production capacity from 29 to 39 pumps.
- Subsequently, the WIP rate, which influences the overall process execution time, was reduced from 15.3 days to 4.5 days. This decrease is the result of the joint implementation of cellular manufacturing and the creation of continuous flow in parallel with customer demand-driven manufacturing.
- Finally, the availability rate increased from 93.90% to 95.98% after the implementation of improvements regarding the creation of a culture of maintenance and early detection of anomalies through two projects 5S and Autonomous Maintenance.

On the environmental front, the improvement in operational performance has significantly improved environmental performance.

- Firstly, specific water consumption has been reduced from 0.6 m3/product to approximately 0.33 m3/product. This is mainly due to a campaign to detect, find and repair water leaks. This campaign was reinforced by the practice of 5S and autonomous maintenance in the machining cell. In addition, the restoration of the basic conditions of the equipment led to adjustments of the pressure and the flow of liquid, in particular for the water widely used in this cell. Other explanations for this result are related to the reduction of the working space obtained by creating the cell manufacturing between the four stations. This led to the reduction of water needed to clean and wash the workshop and storage areas.

- Second, the specific energy consumption was also reduced from 745 Wh/product to 409 Wh/product for several reasons. First, the reduction in product transport and small batch sizes facilitated by the creation of cellular manufacturing and continuous flow. Indeed, product transport consumes after improvements about 9.8% of the total energy used. Second, adjusting the basic equipment conditions has a significant effect on the energy efficiency of the facility.
- Third, the specific consumption of raw metals decreased from 65.17 kg/product to 46.38 kg/product and became closer to the weight of metal in a product which is 40 kg. This was achieved mainly due to the reduction of the defect rate.

It should be noted that the increase in daily production capacity from 29 to 39 pumps has a great effect on the specific consumption of all resources and energy. In addition, the creation of an economy culture promoted by the lean commitment has also further reduced waste in terms of resource use.

Step 3-2: Capitalization and standardization of Lean practices

Having achieved relevant results through several improvements, the team engaged in the establishment of work standards from new ways of doing the activities developed. Standardization of actions has a very powerful impact on sustainability and commitment to perform tasks in accordance with Lean and Green approaches.

The case study shows that the implementation of Lean practices has a very beneficial effect on the environmental performance of SMEs in terms of efficient use of resources such as water, energy and raw materials. Table 3.15 provides information

Table 3.14 Comparison of lean and green metrics before and after improvements implementation

	Metric	Before	After
Productivity and quality metrics	Rate of Time efficiency (%)	49%	62%
	Rate of Quality defect (%)	4.33%	3.61%
	Rate of Inventory (days)	15.3	4.50
	Rate of availability (%)	93.90%	95.98%
Resources and energy consumption metrics	Specific consumption of water (m3/Product)	0.6	0.33
	Specific consumption of Energy (Wh/ product)	745	409
	Specific consumption of metal crude (Kg/ Product)	65.17	46.38

from the case study on how each Lean practice can support companies' environmental management practices in terms of performance improvement.

As explained earlier, lean tools have a positive impact on environmental performance. Practices such as 5S and training can help to spread the culture of eliminating waste of any kind. Value Stream Mapping is a very powerful tool for summarizing information from each process step (in the form of operational and environmental indicators) but also for visualizing anomalies and sources of waste. Also, tools such as cellular manufacturing, SMED, autonomous maintenance contribute to improve productivity, which implies a direct reduction of specific energy and water consumption. Finally, quality control can help reduce production defects, which reduces raw material consumption.

Table 3.15 Impact of lean practices on the green performance in the case of SMEs

Lean practice	Impact on Green performance
5S	- Impacts water and energy consumption through cultural waste eradication - Impacts energy consumption through systematic detection and fixing of waste in energy use
VSM	- Helps to the measurement and the visualization of wastage - Identification of unnecessary operations and opportunity to avoid excess of energy, water and raw material use
Cellular Manufacturing	- Reduces space utilization that minimizes water needed for washing and cleaning storage zone and work area. Also it minimizes the energy use (lightening, handling, conditioning …) - Reduce transport between stations that generates an important source of energy use. Thus it enhances energy efficiency
SMED	- Promotes the continuous flow that reduces energy needed to storage of big lot of product - Helps to produce small lot sizes. This avoids unnecessary use of energy for storage and handling of big lots of products - Reduce non value added time and enhance productivity which reduces the specific consumption of energy and water
Pull system Kanban	- Help reducing WIP in the shop floor with very few batches between the stations - Avoids overproduction that generates inventory and WIP. This avoids unnecessary use of energy for storage and handling of products
Autonomous Maintenance	- Establishes the basic equipment conditions which has a positive impact in reduction of wastage in energy and material - Creates a culture of earlier detection and repair of water leaks. This reduces water consumption - Enhances availability of facilities which avoid use of energy while not producing
Quality control	- Reduces scrap and rework which reduces the specific consumption of raw materials and energy use

3.5 Conclusion

From the case study exposed in this chapter, it is clearly proven that the joint implementation of Lean management and environmental management will only be beneficial for companies. Because the logic of Lean management based on the elimination of all types of waste improves the company's performance, whether it is economic, operational or environmental.

The case study presented here shows that this combination of the two approaches has led to relevant achievements on both the operational and environmental levels without the need for costly investments. In fact, the commitment of the management and the involvement of the staff were the main drivers for the implementation of Lean tools with high added value potential. Managers made the experiences of implementing tools such as 5S, Green VSM, SMED, Kanban, AM and Quality Control, real learning opportunities for employees, which paid off on several levels:

- **Human:** Initiation of a Lean culture among operators and managers.
- **Operational:** Increase of the value-added rate from 49% to about 62%, Decrease of the quality defect rate from 4.33% to 3.61%, The duration of work in progress decreased from 15.3 days to 4.5 days.
- **Environmental:** Reduction of specific water consumption (from 0.6 m3/product to about 0.33 m3/product), Reduction of energy consumption (from 745 Wh/product to 409 Wh/product) and of raw materials (from 65.17 kg/product to 46.38 kg/product).

These results clearly demonstrate the applicability of the Lean and Environmental Management programs in SMEs and the usefulness of this approach to address the internal and external issues facing this type of business. After these results, the challenge is to maintain these achievements and to integrate the environmental component into their daily practices.

References

Agan Y, Acar MF, Borodin A (2013) Drivers of environmental processes and their impact on performance: a study of Turkish SMEs. J Clean Prod 51(15):23–33

Alves JR, Alves JM (2015) Production management model integrating the principles of lean manufacturing and sustainability supported by the cultural transformation of a company. Int J Prod Res 53(17):15

Azzone G, Noci G (1998) Identifying effective PMSs for the deployment of "green" manufacturing strategies. Int J Oper Prod Manag 18(4):308–335

Belhadi A, Touriki FE, Fezazi SE (2016) A framework for effective implementation of lean production in small and medium-sized enterprises. J Ind Eng Manag 9(3):786–810. https://doi.org/10.3926/jiem.1907

Belhadi A, Touriki FE, Elfezazi S (2017) Prioritizing the solutions of lean implementation in SMEs to overcome its barriers: an integrated fuzzy AHP-TOPSIS approach. J Manuf Technol Manag 28(8):1115–1139

Belhadi A, Touriki FE, El Fezazi S (2018a) Lean implementation in small and medium-sized enterprises in less developed countries: some empirical evidence from North Africa. J Small Bus Manag

Belhadi A, Touriki FE, El-Fezazi S (2018b). Benefits of adopting lean production on green performance of SMEs: a case study. Prod Plan & Control

Boiral O, Baron C, Gunnlaugson O (2014) Environmental leadership and consciousness development: a case study among Canadian SMEs. J Bus Ethics 123(3):363–383

Brío JÁ, Junquera B (2003) A review of the literature on environmental innovation management in SMEs: implications for public policies. Technovation 23(12):939–948

Chiarini A (2014) Sustainable manufacturing-greening processes using specific Lean Production tools: an empirical observation from European motorcycle component manufacturers. J Clean Prod 85(15):226–233

Colicchia C, Creazza A, Dallari F (2017) Lean and green supply chain management through intermodal transport: insights from the fast moving consumer goods industry. Prod Plan & Control 28(4):321–334

Cua KO, McKone KE, Schroeder RG (2001) Relationships between implementation of TQM, JIT, and TPM and manufacturing performance. J Oper Manag 19(6):675–694

Drew J, McCallum B, Roggenhofer S (2004) Objectif lean: Réussir l'entreprise au plus juste: enjeux techniques et culturels. Éditions d'Organisation

Dües CM, Tan KH, Lim M (2013) Green as the new Lean: how to use Lean practices as a catalyst to greening your supply chain. J Clean Prod 40:93–100

Faulkner W, Badurdeen F (2014) Sustainable Value Stream Mapping (Sus-VSM): methodology to visualize and assess manufacturing sustainability performance. J Clean Prod 85(15):8–18

Fercoq A (2014) Contribution à la modélisation de l'intégration lean green appliquée au management des déchets pour une performance équilibrée (économique, environnementale, sociale). Thèse, Ecole Natioanale Supérieure d'Arts et Métiers-ENSAM, Paris

Florida R (1996) Lean and green: the move to environmentally conscious manufacturing. Calif Manage Rev 39(1):80–105

Gandhi NS, Thanki SJ, Thakkar JJ (2018) Ranking of drivers for integrated lean-green manufacturing for Indian manufacturing SMEs. J Clean Prod

Garza-Reyes JA, Villarreal B, Kumar V, Ruiz PM (2016) Lean and green in the transport and logistics sector – a case study of simultaneous deployment. Prod Plan & Control 27(15):1221–1232

Gupta V, Narayanamurthy G, Acharya P (2017) Can lean lead to green? Assessment of radial tyre manufacturing processes using system dynamics modelling. Comput & Oper Res.

Hennion R, Makhlouf A (2016) Les fiches Outils du Lean Six Sigma. Eyrolles, Paris

Henry JR (2013) Achieving Lean changeover: putting SMED to work. Taylor & Francis, New York

Imai M 2007. *Gemba Kaizen, l'art de manger avec bon sens.* Kaizen Institute

Jabbour CJ (2013). Environmental management and operational performance in automotive companies in Brazil: the role of human resource management and lean manufacturing. J Clean Prod 57. doi:https://doi.org/10.1016/j.jclepro.2012.07.010

Karim A, Arif-Uz-Zaman K (2013) A methodology for effective implementation of lean strategies and its performance evaluation in manufacturing organizations. Bus Process Manag J 19(1):169–196. https://doi.org/10.1108/14637151311294912

King AA, Lenox MJ (2001) Lean and green? an empirical examination of the relationship between lean production and environmental performance. Prod Oper Manag 10(3):244–256

Liker J (2006) LE MODÈLE TOYOTA: 14 Principes qui feront la réussite de votre entreprise. Village mondial.

Ng R, Low JS, Song B (2015) Integrating and implementing Lean and Green practices based on proposition of carbon-value efficiency metric. J Clean Prod 95(15):242–255

Ohno T (1988) Toyota production system: beyond large-scale production. Productivity Press.

Ortiz C (2006) Kaizen assembly. Taylor and Francis

Pampanelli AB, Found P, Bernardes AM (2014) A lean & green model for a production cell. J Clean Prod 85(15):19–30

Personne M (1998) Contribution à la méthodologie d'intégration de l'environnement dans les PME-PMI: Évaluation des performances environnementales. Thèse, L'institut National des Sciences Appliquées, Lyon,.

Piercy N, Rich N (2015) The relationship between lean operations and sustainable operations. Int J Oper Prod Manag 35(2):282–315

Rahman S, Laosirihongthong T, Sohal AS (2010) Impact of lean strategy on operational performance: a study of Thai manufacturing companies. J Manuf Technol Manag 21(7):839–852. https://doi.org/10.1108/17410381011077946

Ramayah T, Mohamad O, Omar A, Marimuthu M, Leen JY (2012) Green manufacturing practices and performance among SMEs: evidence from. Green Technologies and Business Practices: An IT Approach, pp 208–224

Shah R, Ward PT (2003) Lean manufacturing: context, practice bundles, and performance. J Oper Manag 21(2):129–149

Shah R, Ward PT (2007) Defining and developing measures of lean production. J Oper Manag 25(4):785–805

Sobral MC, Jabbour AB, Jabbour CJ (2013) Green benefits from adopting lean manufacturing: a case study from the automotive sector. Environ Qual Manage 22(3):65–72

Thanki S, Govindan K, Thakkar J (2016) An investigation on lean-green implementation practices in Indian SMEs using analytical hierarchy process (AHP) approach. J Clean Prod

Thanki SJ, Thakkar J (2018) Interdependence analysis of lean-green implementation challenges: a case of Indian SMEs. J Manuf Technol Manag

Vais A, Miron V, Pedersen M, Folke J (2006) "Lean and Green" at a Romanian secondary tissue paper and board mill—putting theory into practice. Resour Conserv Recycl 46(1):44–74

Verrier B, Rose B, Caillaud E (2016) Lean and Green strategy: the Lean and Green House and maturity deployment model. J Clean Prod 116(10):150–156

Vinodh S, Arvind KR, Somanaathan M (2011) Tools and techniques for enabling sustainability through lean initiatives. Clean Technol Environ Policy 13(469–479):3

Womack J, Jones D (2003) Lean thinking: Banish waste and create wealth in your corporation, 2e édition ed. s.l.:Simon & Schuster UK Ltd

Yang MG, Hong P, Modi BS (2010) Impact of lean manufacturing and environmental management on business performance: an empirical study of manufacturing firms. Int J Prod Econ 129(2):251–261

Chapter 4
Lean Six Sigma and Sustainability: From Total Quality to Total Sustainability

Objectives
The current chapter aims to:

1. Provide a glimpse of the evolution of lean six sigma from total quality management to sustainability
2. Highlight how LSS could positively impact sustainability
3. Identify key engines of sustainable LSS
4. Identify the challenges of sustainable LSS

4.1 Introduction

Nowadays, customers, stakeholders and regulators are realizing the importance of dealing with challenges like climate change, pollution, scare resources, high energy cost, poverty and demographical growth (Cherrafi et al. 2016). As a result, high expectations on environmental and social responsibility must be met by today's manufacturers (Touriki et al. 2021). The transformation towards sustainable approaches and strategies is no longer a choice, rather it is an essentiality in order to remain competitive (Garza-Reyes 2015), therefore, companies are called to explore management strategies that would promote the three pillars of sustainability: profit, planet and people. For that purpose, lean six sigma (LSS) is found to be an adequate management strategy that would help manufacturers reach economic performance, while at the same time enhancing the environmental and social aspects (Ali et al. 2020).

Over the years, LSS managed to overcome the shortcomings and weaknesses of total quality management (TQM) (Jourabchi et al. 2014) and successfully provided manufacturers with specific tools and metrics to achieve quality in the ground. It is perceived as the strategy that combines both benefits of lean and six sigma with the aim to improve quality, reduce waste and satisfy customer's needs (Gupta et al. 2018). Today, beyond quality improvements, LSS is also the key answer to the sustainability challenge (Cherrafi et al. 2016). This innovative approach results in cost effective and

F. E. Touriki et al., *Sustainable Excellence in Small and Medium Sized Enterprises*, Industrial Ecology, https://doi.org/10.1007/978-981-19-0371-7_4

sustainable growth, better market position and improved turnover (Kaynak 2003). Moreover, it can enhance green operations, increase awareness of waste and reduce environmental costs (Garza-Reyes 2015). Finally, LSS helps improving health and safety conditions, enhance employee satisfaction and provide them with new skills and expertise (Henao et al. 2019).

In the light of the above, this chapter describes the development of LSS as an approach for sustainability through the elaboration on its impact on the three pillars of the three-bottom line (TBL), the identification of the enablers that would foster success of LSS initiatives for sustainability and the establishment of key challenges that can jeopardize the LSS strategies for sustainability. Critiques that come with the negative impacts of LSS on sustainability are also discussed.

4.2 Background

4.2.1 Lean

Introduced first in Japan, lean manufacturing came to compete with the mass production systems adopted by the USA (Duarte and Cruz-Machado 2019). Today the aim of lean is to produce high quality products at the lowest price possible, in the least time while eliminating waste (Liker and Morgan 2006).

The integrated approach of lean associates employees, suppliers and customers and focuses on eradicating all types of non-value-added operations (Rybkowski et al. 2012). Transport, inventory, motion, waiting, overprocessing, overproduction, and defects are the seven wastes associated with lean manufacturing. Various tools and techniques have been developed to implement lean in manufacturing environments such as Value Stream Mapping (VSM), Single Minute Exchange of Die (SMED), 5S, Kanban, Total Productive Maintenance (TPM)…etc. (Chiarini and Vagnoni 2015). For the successful implementation of lean, all these means must be implemented in a maximum workforce involvement and commitment (Cherrafi et al. 2016).

Lean has been validated by both SMEs and large corporates across all manufacturing sectors not just the automotive sector (Belhadi et al. 2018). Multiple benefits have been documented after the implementation of lean. First, lead times for customers were decreased, inventories for manufacturers were reduced and management knowledge was improved. Second, the manufacturing process has improved since less quality defects and reworks are recorded. Moreover, operational costs are decreased which can deliver significant financial benefits to the company. Finally lean human resources centered approach has helped empowering multi-skilled teams (Melton 2005).

The analysis of several definitions shows a certain complementarity between them but also a divergence of point of view. Some authors consider Lean as a system (Shah and Ward 2007; Hohmann 2012), or process (Womack et al. 1990), others as a philosophy (Liker 2006; Bhasin and Burcher 2006; Karim and Arif-Uz-Zaman 2013), or a set of principles (Drew et al. 2004) (Table 4.1).

In order to facilitate the adoption and understanding of this concept in SMEs, we have constructed a new definition that focuses on the **objective** of Lean, the **means** to achieve it and the **condition** for success. The fact of stating the condition in the definition helps companies to better identify the dimensions of the concept (Operational dimension: production system, human dimension: corporate culture and a managerial dimension: management commitment). The proposed definition is:

> Lean is a systematic and dynamic approach of continuous improvement aiming, through the mobilization of a large array of principles tools and practices, at satisfying the customer with the minimum use of resources.

Table 4.1 Definition of lean according to seven literature reviews

N°	Source	Suggested definition
1	(Womack et al. 1990)	Lean is a dynamic process of change systematically driven by a set of principles and best practices for continuous improvement. Lean combines the positive characteristics of mass production and artisanal production
2	(Liker 2006)	Lean is a philosophy, if implemented, reduces the time between customer order and delivery of the finished product by eliminating sources of waste in the production flow
3	(Drew et al. 2004)	An integrated set of principles, practices, tools and techniques designed to eradicate the causes of poor operational performance, Lean is a systematic approach that aims to eliminate all sources of inefficiency in value chains and close the gap between actual performance and the requirements of customers and shareholders
4	(Bhasin and Burcher 2006)	Lean is a production philosophy, which will be used as a tool to solve problems in a long-term perspective
5	(Shah and Ward 2007)	An integrated socio-technical system whose main objective is to eliminate waste by minimizing, if not eradicating, variability among suppliers, customers, and within the organization
6	(Hohmann 2012)	Lean is a system aimed at generating the maximum added value at the lowest cost and in the shortest time, using the resources just necessary to provide customers with what they value
7	(Karim and Arif-Uz-Zaman 2013)	Lean is a philosophy aimed at reducing waste throughout the company, optimizing core resources and establishing a corporate culture capable of identifying and continuously seeking customer satisfaction

4.2.2 Difference in Perception of Lean Between Developed and Developing Countries

The review of literature on Lean has formally demonstrated that the use and adoption of this approach is more favored and widespread in developed countries such as the United States or the European Union countries Belhadi et al. (2017). Compared to SMEs in developing countries, the implementation initiatives still leave much to be desired and this despite the interest and the proven and approved impact of Lean on industrial performance.

At first glance, we can see that the meaning of Lean implementation is different from one country to another. In the United States, the study carried out by (White et al. 1999) through a survey of 454 companies, showed that the practice of lean is strongly associated with the application of the "just-in-time" pillar. In the same country and through the recent study of (Zhou 2012), it is observed that SMEs are aware of the interest of Lean as a solution that responds to mainly internal issues (reduce costs, improve profit margin, improve equipment availability ...). The companies in this survey have affirmed the practice of the 5S tool, work standardization sheets and quality certifications, as the majority among the Lean tools. Six Sigma is the least used Lean tool, and the authors explain this by the fact that Lean practitioners within SMEs are still novice practitioners and they may not have the capacity to fully implement comprehensive tools such as Six Sigma in their processes.

In Europe, the meaning of Lean is different, its application is mainly focused on the control of process variation through Lean Six Sigma. Furthermore, Thomas et al. (2014) reported that initiatives such as Lean and Six Sigma are being developed and adopted in UK SMEs, although progress is still quite slow as only 4–6% of SMEs have migrated to Lean systems.

In developing countries, such as Thailand, India and Morocco, the application of Lean systems is very recent. It is encouraged by national industrial programs and/or by the installation of large international firms that see them as the ideal place for design, development and manufacturing at a very accessible cost. This suggests that the reasons for the adoption of lean by these SMEs come as a response to mainly external challenges. As evidenced by the study done by (Panizzolo et al. [2012], Lean is considered by these SMEs as a powerful management mode, and which is developed in India only recently. The researchers show that in the beginning the implementation of Lean was difficult and slow, as in other developed countries, due to the slow processes of changing the mindset of people, lack of awareness and training on Lean concepts, the cost and time involved in implementing Lean. India was able to challenge Japanese approaches and means and is now considered among the strong practitioners of Lean production. As a result of the study results, it turns out that the automotive sector is the most excellent in terms of Lean implementation. The main characteristic between the cases studied through this publication, is that all the companies are certified, that it gives special importance to the quality aspect in their organization and manufacturing process.

4.2.3 Six Sigma

Six Sigma was developed at Motorola through the efforts of Bill Smith, a reliability engineer, in the 1980s (Brady and Allen 2006), however, Six Sigma's popularity came through the effort of Jack Welch, the then CEO of General Electric in 1995. Fostered by statistical techniques, it presents a structured and systematic approach to process improvement, aiming for a reduced defect rate. The term "six sigma" itself is used to refer to the statistical measure of defect rate within a system implying that the process output should not be more than 3.4 defects for every million opportunities, or Six Sigma. Six sigma seeks the continuous improvement of the manufacturing process, the elimination of defects root causes and the reduction of process variability.

Six Sigma provides the user with a "more outline of Deming's plan-do-check-act cycle by using a guided five stage cycle called **DMAIC**: define-measure-analyze-improve-control.

– **The define step:**

During this step, we identify the project opportunity, scope, potential and objectives. It is advisable that all parties including customers, stakeholders and the manufacturer agree on a project that is important and useful to all of them.

According to Montgomery and Woodall (2008), several questions should be asked at the end of this step:

- Does the problem statement focus on symptoms, and not on possible causes or solutions?
- Are all the key stakeholders identified?
- Does the potential financial impact make the project worth doing?
- Has the scope of the project been verified to ensure that it is neither too small nor too large?
- Has a high-level process map been completed?
- Have any obvious barriers or obstacles to successful completion of the project been ignored?
- Is the team's action plan for the Measure step of DMAIC reasonable?

– **The measure step:**

The objective of this step is to evaluate quantitively the current state of the process. This includes data collection and the verification of data quality. At this stage, a list of all the key process input variables (KPIV) and the key process output variables (KPOV) should be prepared.

According to Montgomery and Woodall (2008), several questions should be asked at the end of this step:

- There must be a comprehensive process flow chart or value stream map. All major process steps and activities must be identified, along with suppliers and customers.
- A list of KPIVs and KPOVs must be provided, along with identification of how the KPOVs related to customer satisfaction or the customers.

- Measurement systems capability must be documented.
- Any assumptions that were made during data collection must be noted.
- The team should be able to respond to requests such as, "Explain where that data came from" and questions such as, "How did you decide what data to collect?", "How valid is your measurement system?", and "Did you collect enough data to provide a reasonable estimate of process performance?"

– **The analyze step:**

The objective of this step is to determine the cause-and-effect relationships in the process and to understand the different sources of variability. The purpose is to highlight the potential causes of the defects, quality problems, customer issues and problems that motivated the project.

According to Montgomery and Woodall (2008), several questions should be asked at the end of this step:

- What opportunities are going to be targeted for investigation in the Improve step?
- What data and analysis support that investigating the targeted opportunities and improving/eliminating them will have the desired outcome on the KPOVs and customer CTQs that were the original focus of the project?
- Are there other opportunities that are not going to be further evaluated? If so, why?
- Is the project still on track with respect to time and anticipated outcomes?
- Are any additional resources required?

– **The Improve step:**

In the Improve step, the team must come up with specific changes and solutions that can be made in the process in order to tackle the identified issues, remove their causes and have the desired impact on process performance.

According to Montgomery and Woodall (2008), several questions should be asked at the end of this step:

- Adequate documentation of how the problem solution was obtained.
- Documentation on alternative solutions that were considered.
- Complete analysis results for the pilot test.
- Plans to implement the pilot test results on a full-scale basis.
- Analysis of any risks of implementing the solution, and appropriate risk-management plans

– **The control step:**

The objective of this stage is to provide a process control plan and the requiered procedures to ensure gains capitalization and standardization, in a way to make sure that the improvements will be implemented in other similar processes in the business.

According to Montgomery and Woodall (2008), several questions should be asked at the end of this step:

- Data illustrating that the before and after results are in line with the project charter should be available. (Were the original objectives accomplished?)
- Is the process control plan complete? Are procedures to monitor the process, such as control charts, in place?
- Is all essential documentation for the process owner complete?
- A summary of lessons learned from the project should be available.
- A list of opportunities that were not pursued in the project should be prepared.
- A list of opportunities where the results of the project can be used in other parts of the business.

In each stage a variety of tools can be used in order to measure, analyze and improve processes with the aim to reduce variability and lead the system to a state of control. Table 4.2 provides some examples of the statistical techniques that can be used during each step of the DMAIC cycle.

Across the literature, there is a variety of advantages following the use of six sigma. Eight main characteristics are behind the incredible success of six sigma. While other quality methodologies only check one or two of these characteristics, six sigma fills them all.

(i) bottom-line results expected and delivered.
(ii) senior management leadership.
(iii) a disciplined approach (i.e. DMAIC).
(iv) rapid (3–6 month) project completion.
(v) clearly defined measures of success.
(vi) infrastructure roles for six sigma practitioners and leadership.
(vii) focus on customers and processes; and
(viii) a sound statistical approach to improvement.

Table 4.2 Example of statistical tools used in DMAIC (Requeijo et al., 2008)

	Define	Measure	Analyze	Improve	Control
Project charter	x				
Process maps & flow charts	x	x			
Cause and effect analysis		x			
Process capability analysis		x			
Hypothesis tests, confidence intervals			x		
Regression analysis, other multivariate methods			x		
Gauge R&R		x			
Failure mode & effects analysis			x		
Designed experiments			x	x	
SPC and process control plans		x	x		x

4.2.4 Lean Six Sigma

LSS can be defined as a structured and systematic approach that aims to maximize shareholder value through the improvement of quality, lead times, customer satisfaction and costs (Laureani and Antony 2012). It is a new generation that combines the techniques and principles of both lean and six sigma. According to Wheat, Mills, and Carnell (2003), six sigma supplements lean by offering a set of tools and "know-how" to deal with manufacturing problems that have been identified through lean.

Companies in various industries have realized the importance of LSS as a leading continuous improvement technique to overcome the shortcomings of both lean and six sigma, as deploying six sigma in isolation is not enough to remove all kinds of waste and implementing lean in isolation cannot remove variation from the process (Corbett 2011).

Based on the Table 4.3, we can conclude that lean and Six Sigma are perfectly suited to be used in a comprehensive methodology that integrates both elements. LSS targets different opportunities for improvement, while six sigma only targets specific individuals and scopes. Lean on the other hand empowers and engages all the personnel in the organization. Therefore, Six Sigma could drive the implementation of lean efforts and initiatives in an improvement environment, as it can be challenging to gain momentum when attempting to extend the philosophy. Consequently, efforts need to be directed by a strong approach that can preserve direction and focus within the organization.

A successful implementation of LSS results in process optimization and efficiency, improved product quality, improved competitive advantage, increased bottom

Table 4.3 Similarities between lean and six sigma

Lean	Six Sigma
Establish methodology for improvement	Policy deployment methodology
Focus on customer value stream	Customer requirements measurement, cross functional management
Use a project-based implementation	Project management skills
Understand current conditions	Knowledge discovery
Collect product and production data	Data collection and analysis tools
Document current layout and flow	Process mapping and flowcharting
Time the process	Data collection tools and techniques, SPC
Calculate process capacity and Takt time	Data collection tools and techniques, SPC
Create standard work combination sheets	Process control planning
Evaluate the options	Cause-and-effect, FMEA
Plan new layouts	Team skills, project management
Test to confirm improvement	Statistical methods for valid comparison, SPC
Reduce cycle times, product defects, changeover time, equipment failures, etc	Seven management tools, seven quality control tools, design of experiments

line and sustainability, decreased customer returns, increased production capacity, reduced costs, better visibility of variation and waste in the process, improved customer satisfaction and reduced cycle and lead times (Albliwi et al. 2015).

Across the literature, key elements that were deemed to be necessary for the successful development of LSS initiatives are as follow:

- Committed leadership
- Talented people
- Supporting infrastructure
- Project selection, prioritization, reporting and tracking system
- Culture change
- Access to sufficient data
- Sufficient personnel to carry out projects
- Know-how of tools and techniques
- LSS roadmap

SMEs can greatly benefit from the use of LSS. Antony et al. (2005) have conducted a a multi-case study within four SMEs in the United Kingdome; where he compared the firms using six sigma vs the ones that do not apply it. The reported results showed that Six Sigma SMEs realized a significant improvement in scrap rate, throughput yield, cycle time, on-time delivery as well as sales, profit margin, customer satisfaction compared to enterprises that are not deploying LSS. Another study conducted by Thomas et al. (2014) have highlighted the several improvements in business performance following the implementation of LSS in SME. For example, the authors noted a reduction in build time, on time in full delivery and a reduction of the non-value-added time. Furthermore, important financial savings were recorded from this case study.

- The implementation of LSS in SMEs should consider the characterization, compatibility and innovativeness of SMEs toward LSS implementation. This implementation, also requires SMEs to develop and improve themselves by showing the following characteristics
- Stronger and committed leadership
- Visible management commitment
- Stronger financial capabilities
- Enhanced skills and expertise of the personnel
- Conducive organizational culture.
- Clear definition of customer requirements
- Shared understanding of core business processes and their critical characteristics
- Rewarding and recognizing the team members
- Communicating the success and failure stories

4.2.5 *Total Quality Management*

TQM can be defined as a management philosophy and culture that seeks customer satisfaction involving all employees in continual improvement (Arditi and Gunaydin 1997). It uses data and effective communication in order to incorporate quality in the organization's culture and activities. Throughout the literature, researchers agree on the main requirements of TQM to be customer focus, leadership, quality planning, factual management, continuous improvement, human resource management, learning, process management, supplier's relationships and organizational awareness of the social and environmental context (Claver et al. 2003). TQM includes various tools and techniques for quality improvement that can be classified into two categories "soft tools" such as run charts, Pareto, brainstorming, flow charts, etc. and "hard tools" that deals with production and work process control techniques like process design, the "just in time" philosophy, the norm ISO 9000 and the basic quality control tools (Wilkinson 1998).

TQM allows firms to strengthen their competitive position, enhance their productivity and market image and improve their customer focus and satisfaction (Kanji 1998; Gunasekaram et al. 1998), however, this philosophy has been heavily criticized for the fact that it is only a "philosophy" that need to move beyond this view (Pepper et al. 2010).

4.3 Lean Six Sigma: From Total Quality Management to Sustainable Development

Manufacturers have realized that they need more comprehensive approaches than prior quality initiatives like TQM (Kwak and Anbari 2006). TQM emerged as an all-inclusive philosophy that offers a rich set of quality concepts and principles; however, firms need to move beyond this and translate these theorical concepts into methodologies that can achieve process improvements (Zu et al. 2008).

From a historical perspective, since 1980, there has been a significant growth in the use of statistical methods for quality and overall business improvement throughout the world. This has enhanced firms' competitiveness with respect to quality. In this perspective, several methodologies and management systems have emerged aiming to implement quality initiatives. TQM was one of these management strategies for managing quality improvement activities. TQM was mainly influenced by the philosophies of W. Edwards Deming and Joseph Juran. TQM, afterwards integrated a variety of concepts like participative organizations, work culture, customer focus and supplier quality improvement. However, little to moderate success was registered with the implementation of TQM, as there no link nor integration between the quality system and the organization goals, in addition to the lack of technical tools and the absence of recognition of the importance of variability reduction.

Across the literature, there is a general agreement that six sigma is more successful than TQM. This is mainly due to the focus of SS on projects that directly impact the financial performance and provide real savings to the company, besides customer satisfaction.

Indeed, both TQM and Six Sigma share the common aim of reducing the cost of poor quality and increase customer satisfaction. They both involve staff and need their commitment and motivation. Nevertheless, six sigma is a more structural and hierarchical methodology that uses DMAIC that is suitable for important and long-term improvements. According to x, most tools of six sigma are derived from TQM that is quite a philosophy than an exact methodology.

At the same time, lean philosophy emerged as integrated approach that associates employees, suppliers and customers and focuses on eradicating all types of non-value-added operations (Rybkowski et al. 2012). However, despite the significant advantages of lean and SS, both of them are escorted with several shortcomings that encouraged manufacturers to marry both methodologies around the year of 2000. This marriage was was possible owing to the similarities shared between the two methodologies, as both require an adequate culture of continuous improvement. LSS has managed to overcome the weaknesses of TQM and associated quality with specific metrics that enable manufacturers to quantify quality improvements through SS and at the same time empower and engage all the levels of the organization through lean (Kwak and Anbari 2006).

The evolution of LSS is presented in Fig. 4.1 (Ruben et al. [2018]).

Nevertheless, today's manufacturers no longer rely solely on quality; sustainability has emerged to be a part to the equation. The emphasis on the concept of sustainability in businesses and operations is increasing with the growing pressure from stakeholders, government and public organizations (Ali et al. 2020). As a result, companies are forced to shift towards sustainable approaches and strategies (Garza-Reyes 2015).

According to Cherrafi et al. (2016), LSS answers the challenge of sustainability, therefore it moves from being a just a methodology that reinforces TQM (Zu et al. 2008), to a management system that supports the three pillars of sustainability. Recent literature argued that the integration between lean, six sigma and sustainability is a

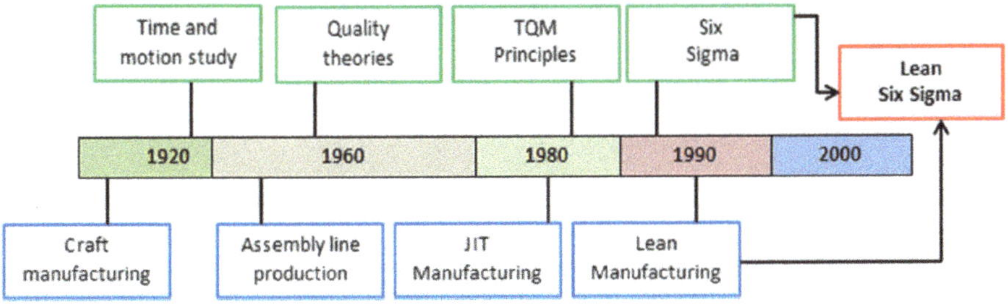

Fig. 4.1 The evolution of lean six sigma (Ruben, Vinodh, and Asokan 2018)

synergetic operation, since these principles align together in order to achieve operational excellence while enhancing the bottom-line results. LSS, therefore is the solution to meet stakeholders and customer's need, attain positive economic performance and balance sustainability pillars (Alves and Alves 2015) (Cherrafi et al. 2016; Hajmohammad et al. 2013; Martínez-Jurado and Moyano-Fuentes 2014). In a nutshell, sustainability is no longer a luxury or a "nice to have" rather it has become a business imperative that can be reached through the LSS concepts (Garza-Reyes 2015).

4.4 Lean Six Sigma: Towards Sustainability

4.4.1 LSS and Economic Performance

Many researchers have investigated the impact of LSS on economic and operational performance. As a new generation of quality improvement, LSS results in fewer defects and rework, lower levels of inventory, faster production, less space requirement, less transportation, less waiting, more efficient resource utilization which leads to an enhanced business performance and increased return on assets through cost reduction (Bortolotti et al. 2018; Sajan et al. 2017; Martínez-Jurado and Moyano-Fuentes 2014) argued that LSS practices and tools are linked to financial performance, besides, they facilitate cost effective and sustainable growth of the company. Moreover, LSS can improve the company's reputation which consequently improves the market position, competitiveness and turnover (Kaynak 2003).

The study of Hill et al. (2018) highlighted the positive impact of six sigma on the enhancement of maintenance and repair which can significantly decrease operational cost. (Snee 2010) pinpointed that companies registered a return of 1 to 4 percent of sales per year upon the implementation of LSS projects. Furthermore, Ertürk et al. (2016) recorded an 80 to 100% increase in business profitability due to the LSS employment.

Kumar et al. (2006) also elaborated on how LSS practices in Indian SMEs has resulted in machine downtime and inventories which directly impact operational performance and costs.

In a nutshell, LSS can maximize and sustain business profitability, meet customer needs and satisfaction, and reduce operational costs, therefore, it positively impacts the economic pillar of sustainability.

4.4.2 LSS and Environmental Performance

Literature has largely elaborated on the benefits of LSS on environmental performance.

Garza-Reyes (2015) argued that LSS can enhance green operation, increase the awareness of waste and provide employees with tools to understand causes and effects of environmental issues. Fatemi and Franchetti (2016) asserted that LSS application can result in significant environmental cost savings; De Oliveira Furukawa et al. (2016) registered a considerable reduction in packaging, plastic bags and material disposals post the implementation of LSS. Sony, Naik, and Antony (2020) and Al-Shamkhani and Elshennawy (2013) also specified respectively on how pollution prevention programs, environmental impacts of discharging oils were improved after LSS projects initiations.

The U.S Environmental Protection Agency (EPA) also supports the existing literature through several studies that have been conducted in American public sectors in order to examine the positive impacts of LSS on the environment. The results indicate that LSS approach can be beneficial to both operational and environmental performance; therefore, LSS can create a competitive business that is also eco-friendly.

In a nutshell, as manufacturing systems are becoming environmentally sustainable, LSS holds in its sleeves great potential to enable environmental and operational benefits (Ruben et al. 2018).

4.4.3 LSS and Social Performance

Although, there is little literature discussing the impact of LSS on the social dimension of sustainability, most agree on the relatively positive outcomes of LSS socially. Erdil, Aktas, and Arani (2018) argued that it is important to embrace all the three dimensions of sustainability, including the social one, when discussing LSS.

In the LSS environment, health and safety conditions are improved, employee satisfaction is increased, and new skills are acquired (Henao et al. 2019; Sakthi Nagaraj and Jeyapaul 2020; Sony and Naik 2020). Lean practices could provide a safer and comfortable shop floor environment where workers are satisfied and comfortable in their working environment. Moreover, the lean approach was found to improve teamwork and connect people, besides companies that deploy lean systems can accept green or social ambitions easier. Furthermore, LSS has a positive impact on the employees personal and professional skills, since LSS approaches require a high degree of training and effort (Aboelmaged 2011). Into the bargain, LSS encourages the implication of management at every stage, which also increases the employee's satisfaction since they feel more recognition from management. Finally, Thomas et al. (2008), found that the rejection rate of new management initiatives has decreased after the implementation of LSS, consequently employees start to sense a higher degree of accomplishment.

However, LSS require standardization which can result in routinely operations, lack of motivation, slippage back to old behaviors and even sabotages of the LSS projects (Ketokivi and Schroeder 2004). In addition, low level of empowerment and lack of management appreciation can discourage workers and jeopardize LSS efforts. Therefore, management involvement is highly required to drive change from the bottom floor, show visible commitment, appreciation and encouragement and actively participate in LSS events (Worley and Doolen 2015).

In a nutshell, we can conclude that LSS can positively impact the social sustainability only if critical conditions are reached.

4.5 Key Engines of Sustainable Lean Six Sigma

The identification of enablers for the success of LSS initiatives for sustainability is highly important in order to help companies focus their effort on these factors to increase their chances of success. Throughout the literature, there are number of critical factors that leads to the fruitful implementation of LSS for reaching sustainability.

4.5.1 Continuous Improvement and Innovation Culture

The LSS philosophy needs a culture of continuous improvement where the company should continuously focus on waste elimination. All employees need to be involved across the organization in order to identify opportunities for improvement, investigate root causes of problems and search for ways to eliminate waste (Cherrafi et al. 2019).The process of attaining sustainability is an everyday-day effort. The creation of a continuous improvement and supportive LSS culture where values such as collaboration, transparency and openness are keys to the development of sustainable LSS. In a case study conducted by Caldera, Desha, and Dawes (2017), respondents identified continuous improvement tools as key enablers that helped them increase their leanness ranging from 5S, PDCA to VSM, thus catalyzing the sustainability of LSS projects. Hence companies need to develop an organizational culture that promote improvement initiatives across the organization (Dahlgaard and Dahlgaard-Park 2006). On the other hand, innovation acts as a catalyst for the company's competitiveness through sustainable LSS. Innovation provides manufacturers with solutions to business, environmental and social challenges (Hult et al. 2004). An innovative culture can foster employee's creativity and skills which further enhance their problem-solving capabilities. This can enhance employee's self-esteem and also improve their productivity and effectiveness.

4.5.2 *Management Involvement and Leadership*

There is a mutual agreement across the literature about the crucial role of management involvement and leadership in promoting sustainable LSS (Paranitharan and Jeyathilagar 2017) that both visionary leadership and management commitment result in higher performance. Into the bargain, it was found that managers who well-articulate guidelines and show a high level of motivation and commitment are more likely to drive continuous changes for sustainability (Alves and Alves 2015); under their management style, employees feel more recognized and motivated to give more. However, managers need to be careful not to confuse involvement with micromanagement that can hinder the long-term sustainability of LSS success (Gelei et al. 2015).

4.5.3 *People's Involvement and Training*

A well trained and motivated staff with adequate skills and expertise is of upmost importance in order to succeed LSS for sustainability. In the environment of LSS, employees need to be knowledgeable of the principles that make LSS, in addition it is the company's responsibility to provide them with the necessary tools and training in order to succeed the LSS projects. Hence the continuation of training programs on LSS and the boost of employee's skills is critical to develop an LSS expertise.

4.5.4 *Effective Communication*

Effective top-down communication was found to be a requirement for sustainable LSS, under which employees feel valued and recognized. This can foster a culture of exchange, growth and continuous improvement (Jeyaraman and Teo 2010). Efficient and frequent communication is key to the success of LSS towards sustainability which can provide employees with regular instructions to preserve the momentum and increase their efforts to reach continuous improvements. Therefore, good communication is a key ingredient for the success of LSS initiatives and sustainability.

4.5.5 *Stakeholders' Involvement*

The successful adoption of sustainable LSS must include all stakeholders across the organization. The involvement of these stakeholders can increase the staff empowerment and responsibility (Wahab et al. 2013). Moreover, suppliers' integration and customer satisfaction are one of the most important pieces of LSS implementation, therefore higher performance can be achieved through the alliance of all stakeholders (Paranitharan and Ramesh Babu 2014).

4.5.6 *IT and Information Systems*

The exploitation of new advanced technologies is crucial in today's manufacturing environment (Paranitharan and Jeyathilagar 2017). The emerging technologies of the industry 4.0 like big data, IOT and smart factories have an enormous impact on the sustainable business performance in the SMEs. Haseeb, M., Hussain, H. I., Ślusarczyk, B., & Jermsittiparsert, K. highlights that managers will have greater sustainable business performance when implementing IoT.

Key applications of industry 4.0 are smart factories that result in the manufacturing of a variety of smart products leading to a positive impact on SBE. Hence, we can only conclude that the better implementation of IT has a major role in boosting SBE.

4.6 Sustainable Lean Six Sigma: Challenges Ahead

The implementation of LSS for sustainability journey is not easy, several challenges need to be addressed in order to succeed. Resistance to change, negative organizational values, lack of environmental and social awareness, inadequate management capacity, lack of environmental and social responsibility and the lack of human resources involvement in the LSS projects are among the organizational barriers that can hinder the opportunities of achieving sustainable developments and benefits (Cherrafi et al. 2016; Abdulsaleh and Worthington 2013; Dakov and Novkov 2007). Manufacturers will consider integrating the environmental and social aspect of the TBL if financial benefits can be made. This lack of awareness is also justified by the traditional perspective of considering environmental and social performance as a hurdle to economic performance. Moreover, employee's resistance to adopt new sustainable approaches of LSS can be explained by their fear of changing their routinely habits and tackling a relatively novel concept. Furthermore, the lack of confidence and knowledge of perceived benefits among managers stop them from adopting sustainable LSS projects.

Unsupportive organizational culture can also be an obstacle to the sustainable LSS implementation, under this culture, management fails to involve their employees, fails to support and actively participate in LSS projects and fails to set a role model for other employees (Kumar et al. 2006). On the other hand, sustainable LSS projects require financial resources. Sustainable LSS brings paradigm shift in the manufacturing process, therefore these changes need to be financed (Kaswan and Rathi 2020; Henriques and Catarino 2016). Besides, LSS call for employee's extensive training which comes at a high cost, time and require external assistance and expertise (Belhadi et al. 2018).

However, according to (Cherrafi et al. 2016), the implementation of LSS and sustainability can soothe the identified challenges through the reduction of sustainability initiatives costs, the enhancement of awareness across the organization, the development of an organizational culture that is highly favorable to sustainable practices. As a result, manufacturers will not only improve their operational performance but also create a conducive environment of sustainable LSS.

4.7 Is LSS Always Sustainable?

Despite the rich literature about the positive role of LSS in promoting sustainability, there are many criticism topics that suggests that LSS is not always sustainable.

According to Hartini and Ciptomulyono (2015), not all lean principles are positively related to environmental performance, isolated lean practices cannot tackle all the environmental issues. One example is the just in time practice that might increase CO2 emissions in the case of small lot delivery orders or in the case of uncommitted suppliers (Khodeir and Othman 2018).

On the social level, lean create a stressful, uncomfortable environment since lean requires prompt responses to customers' requirements. Moreover, Pepper et al. (2010) labels six sigma as a heavy technical approach which can be viewed as a complex approach that is difficult to understand and be applicable by all employees. Besides, authors like Antony (2004) suggest that six sigma can shift the business into a relatively "bureaucratic" and time-consuming process that oppresses creativity and turn employees into standardized robots, if the focus is on secondary aspects such as number of trained black or green belts instead of bottom-line impact.

Nevertheless, Khodeir and Othman (2018) suggest that LSS and sustainability principles could be promoted at the same time in certain conditions, for example, the just in time practice can support sustainability in case of large deliveries and committed suppliers. They also assumed that lean can be considered sustainable only when customers orientations and values are sustainable.

As a result, it can be concluded that LSS can have a clear positive impact on all aspects of sustainability only if certain conditions are met; critical success factors are followed, and challenges are overcome.

4.8 Conclusion

LSS identifies as a successful management tool that aims to improve product quality, reduce operational costs and boost customer satisfaction. From quality theories, to TQM, LSS was born to overcome the shortcomings of both lean manufacturing and six sigma and provide manufacturers with practical insights to solve quality problems and respond to customer needs.

LSS evolved to not only tackle quality problems but also address sustainability challenges. Responding to regulatory pressures for environmental sustainability and social responsibility, complying with stakeholders' requirements and keeping up with public opinion shifted the focus of both scholars and practitioners to investigate how can LSS help reaching sustainability goals. In this chapter we elaborate on the impact of LSS on the three pillars of sustainability (profit, planet, people). Furthermore, we identified a set of five enablers for the success of LSS initiatives for sustainability. Moreover, we highlighted the challenges and barriers that can hinder this implementation if not addressed properly. Finally, in order to give an objective overview of LSS and sustainability we discussed the critiques that surrounds LSS and how it is not always sustainable.

References

Abdulsaleh, Abdulaziz M, Andrew C Worthington (2013) Small and medium-sized enterprises financing: a review of literature. Int J Bus Manag 8(14). Citeseer, 36

Aboelmaged MG (2011) Reconstructing six sigma barriers in manufacturing and service organizations: the effects of organizational parameters. Int J Qual Reliab Manag 28(5):519–541. https://doi.org/10.1108/02656711111132562

Albliwi SA, Antony J, Abdul S, halim (2015) A systematic review of lean six sigma for the manufacturing industry. Emerald Group Publishing Limited, Bus Process Manag J

Ali, Yousaf, Ahsan Younus, Amin Ullah Khan, Hamza Pervez (2020) Impact of lean, six sigma and environmental sustainability on the performance of SMEs. Int J Prod Perform ManagementBus Manag Stud. Emerald Publishing Limited

Al-Shamkhani, Maher, Ahmed K Elshennawy (2013) Application of six sigma methodology to reduce environmental and economic impacts of discharging the produced water in Southern Iraqi oil fields. J Manag & Eng Integr 6(2). International Conference on Industry, Engineering, & Management Systems, 31

Alves, José Roberto Xavier, João Murta Alves (2015) Production management model integrating the principles of lean manufacturing and sustainability supported by the cultural transformation of a company. Int J Prod Res 53(17). Taylor & Francis: 5320–5333

Antony J (2004) Some pros and cons of six sigma: an academic perspective. TQM Mag 16(4):303–306. https://doi.org/10.1108/09544780410541945

Antony J, Kumar M, Madu CN (2005) Six sigma in small-and medium-sized UK manufacturing enterprises: Some empirical observations. Int J Qual Reliab Manag

Arditi, David, H Murat Gunaydin (1997) Total Quality Management in the construction process. Int J Proj Manag 15(4). Elsevier, 235–243

Belhadi, A, Touriki, FE, El-Fezazi, S (2017) Lean implementation in small and medium-enterprises in less developed countries: some empirical evidences from North Africa. J Small Bus Manag

Belhadi, Amine, Yusof Bin Mohd Sha'ri, Fatima Ezahra Touriki, Said El Fezazi (2018) Lean production in SMEs: literature review and reflection on future challenges. J Ind Prod Eng 35(6). Taylor & Francis, 368–382

Bhasin S, Burcher P (2006) Lean viewed as a philosophy. J Manuf Technol Manag 17(1):56–72. https://doi.org/10.1108/17410380610639506

Borim-de-Souza R, Balbinot Z, Travis EF, Munck L, Takahashi ARW (2015) Sustainable development and sustainability as study objects for comparative management theory: proposing styles of reasoning for an unknown metropole. Emerald Group Publishing Limited, Cross Cultural Management

Bortolotti T, Boscari S, Danese P, Suni HAM, Rich N, Romano P (2018) The social benefits of kaizen initiatives in healthcare: an empirical study. Int J Oper Prod Manag 38(2):554–578. https://doi.org/10.1108/IJOPM-02-2017-0085

Brady, James E, Theodore T Allen (2006) Six sigma literature: a review and agenda for future research. Qual Reliab Eng Int 22(3). Wiley Online Library, 335–367

Caldera, HTS, Cheryl Desha, Les Dawes (2017) Exploring the role of lean thinking in sustainable business practice: a systematic literature review. J Clean Prod 167. Elsevier, 1546–1565

Cherrafi, Anass, Said Elfezazi, Andrea Chiarini, Ahmed Mokhlis, Khalid Benhida (2016) The integration of lean manufacturing, six sigma and sustainability: a literature review and future research directions for developing a specific model. J Clean Prod 139. Elsevier, 828–846

Cherrafi, Anass, Said Elfezazi, Brion Hurley, Jose Arturo Garza-Reyes, Vikas Kumar, Anthony Anosike, Luciano Batista (2019) Green and lean: a Gemba-Kaizen model for sustainability enhancement. Prod Plan Control 30(5–6). Taylor & Francis, 385–399. https://doi.org/10.1080/09537287.2018.1501808

Chiarini, Andrea, Emidia Vagnoni (2015) World-class manufacturing by fiat. Comparison with Toyota Production System from a strategic management, management accounting, operations management and performance measurement dimension. Int J Prod Res 53(2). Taylor & Francis, 590–606

Claver, Enrique, Juan Jose Tari, José Francisco Molina (2003) Critical factors and results of quality management: an empirical study. Total Qual Manag & Bus Excel 14(1). Taylor & Francis, 91–118

Corbett, Lawrence M (2011) Lean six sigma: the contribution to business excellence. Int J Lean Six Sigma. Emerald Group Publishing Limited

Dahlgaard JJ, Dahlgaard-Park SM (2006) Lean production, six sigma quality, TQM and company culture. TQM Mag 18(3):263–281. https://doi.org/10.1108/09544780610659998

Dakov, Ivan, Svetoslav Novkov (2007) Assessment of the lean production effect on the sustainable industrial enterprise development. Bus: Theory Pract 8(4): 183–188. https://doi.org/10.3846/btp.2007.26

Drew, J, McCallum, B, Roggenhofer, S (2004) Objectif lean: Réussir l'entreprise au plus juste: enjeux techniques et culturels. Éditions d'Organisation

Duarte, Susana, V Cruz-Machado (2019) Green and lean supply-chain transformation: a roadmap. Prod Plan & Control 30(14). Taylor & Francis, 1170–1183

Elkington, John (1994) Towards the sustainable corporation: win-win-win business strategies for sustainable development. Calif Manag Rev 36(2). SAGE Publications Sage CA: Los Angeles, CA, 90–100

Erdil NO, Aktas CB, Arani OM (2018) Embedding sustainability in lean six sigma efforts. J Clean Prod 198:520–529. https://doi.org/10.1016/j.jclepro.2018.07.048

Ertürk, Muzaffer, Muhemmetali. Tuerdi(Maimaitiaili. Tuerdi), Aihemaituoheti Wujiabudula (2016) The effects of six sigma approach on business performance: a study of white goods (home appliances) sector in Turkey. Procedia - Soc Behav Sci 229:444–452.https://doi.org/10.1016/j.sbspro.2016.07.154

Fatemi, Shila, Matthew J Franchetti (2016) An application of sustainable lean and green strategy with a six sigma approach on a manufacturing system. Int J Six Sigma Compet Advant 10(1). Inderscience Publishers (IEL), 62–75

Garza-Reyes JA (2015) Green lean and the need for six sigma. Int J Lean Six Sigma 6(3):226–248. https://doi.org/10.1108/IJLSS-04-2014-0010

Gelei A, Losonci D, Matyusz Z (2015) Lean production and leadership attributes - the case of Hungarian Production Managers. J Manuf Technol Manag 26(4):477–500. https://doi.org/10.1108/JMTM-05-2013-0059

Gunasekaram, A, Suresh K Goyal, Teppo MArtikainen, Paavo Yli-Olli (1998) Total Quality Management: a new perspective for improving quality and productivity. Int J Qual & Reliab Manag. MCB UP Ltd

Gupta, Vikash, Rahul Jain, ML Meena, Dangayach GS (2018) Six-sigma application in tire-manufacturing company: a case study. J Ind Eng Int 14(3). Springer, 511–520

Hajmohammad S, Vachon S, Klassen RD, Gavronski I (2013) Lean management and supply management: their role in green practices and performance. J Clean Prod 39:312–320. https://doi.org/10.1016/j.jclepro.2012.07.028

Hartini, Sri, Udisubakti Ciptomulyono (2015) The relationship between lean and sustainable manufacturing on performance: literature review. Procedia Manuf 4. Elsevier, 38–45

Henao, Rafael, William Sarache, Iván Gómez (2019) Lean manufacturing and sustainable performance: trends and future challenges. J Clean Prod 208. Elsevier, 99–116

Henriques J, Catarino J (2016) Motivating towards energy efficiency in small and medium enterprises. J Clean Prod 139(2016):42–50. https://doi.org/10.1016/j.jclepro.2016.08.026

Hill, J, Thomas, AJ, Mason-Jones, RK, El-Kateb, S (2018) The implementation of a lean six sigma framework to enhance operational performance in an MRO facility. Prod Manuf Res 6(1). Taylor & Francis, 26–48. https://doi.org/10.1080/21693277.2017.1417179

Hohmann, C (2012) Lean management: outils, méthodes, retours d'expériences, questions/réponses. Editions d'Organisation

Hult, G, Thomas M, Robert F Hurley, Gary A Knight (2004) Innovativeness: its antecedents and impact on business performance. Ind Mark Manag 33(5):429–438.https://doi.org/10.1016/j.indmarman.2003.08.015

Jeyaraman K, Teo LK (2010) A conceptual framework for critical success factors of lean six sigma: implementation on the performance of electronic manufacturing service industry. Int J Lean Six Sigma 1(3):191–215. https://doi.org/10.1108/20401461011075008

Jourabchi, SM Mehdi, Tina Arabian, Z Leman, Md Yusof B Ismail (2014) Contribution of lean and six sigma to effective cost of quality management. Int J Prod Qual Manag 14(2). Inderscience Publishers Ltd, 149–165

Kanji, Gopal K (1998) Measurement of business excellence. Total Quality Management 9(7). Taylor & Francis, 633–643

Karim A, Arif-Uz-Zaman K (2013) A methodology for effective implementation of lean strategies and its performance evaluation in manufacturing organizations. Bus Process Manag J 19(1):169–196. https://doi.org/10.1108/14637151311294912

Kaswan, Mahender Singh, Rajeev Rathi (2020) Green lean six sigma for sustainable development: integration and framework. Environ Impact Assess Rev 83 (November 2019). Elsevier, 106396. https://doi.org/10.1016/j.eiar.2020.106396

Kaynak, Hale (2003) The relationship between Total Quality Management practices and their effects on firm performance. J Oper Manag 21(4). Elsevier, 405–435

Ketokivi MA, Schroeder RG (2004) Perceptual measures of performance: fact or fiction? J Oper Manag 22(3):247–264. https://doi.org/10.1016/j.jom.2002.07.001

Khodeir, Laila M, Reem Othman (2018) Examining the interaction between lean and sustainability principles in the management process of AEC industry. Ain Shams Eng J 9(4). Ain Shams University, 1627–1634. https://doi.org/10.1016/j.asej.2016.12.005

Kumar M, Antony J, Singh RK, Tiwari MK, Perry D (2006) Implementing the lean sigma framework in an Indian SME: a case study. Production Planning and Control 17(4):407–423. https://doi.org/10.1080/09537280500483350

Kwak, Young Hoon, Frank T Anbari (2006) Benefits, obstacles, and future of six sigma approach. Technovation 26(5–6). Elsevier, 708–715

Laureani, Alessandro, Jiju Antony (2012) Standards for lean six sigma certification. Int J Prod Perform Manag. Emerald Group Publishing Limited

Liker, J (2006) LE MODÈLE TOYOTA: 14 Principes qui feront la réussite de votre entreprise. Village mondial

Liker, Jeffrey K, James M Morgan (2006) The Toyota way in services: the case of lean product development. Acad Manag Perspect 20(2). Academy of Management Briarcliff Manor, NY 10510, 5–20

Magnusson, Kjell, Dag Kroslid, Bo Bergman, Peter Häyhänen, and Don Mills (2003) Six sigma: the pragmatic approach. Studentlitteratur

Martínez-Jurado, Pedro José, José Moyano-Fuentes (2014) Lean management, supply chain management and sustainability: a literature review. J Clean Prod 85. Elsevier Ltd, 134–150. https://doi.org/10.1016/j.jclepro.2013.09.042

Melton, Trish (2005) The benefits of lean manufacturing: what lean thinking has to offer the process industries. Chem Eng Res Des 83(6). Elsevier, 662–673

Missimer, Merlina (2015) Social sustainability within the framework for strategic sustainable development. Blekinge Tekniska Högskola

Montgomery DC, Woodall WH (2008) An overview of six sigma. Int Stat Rev/Revue Internationale de Statistique, 329–346

Morelli J (2011) Environmental sustainability: a definition for environmental professionals. J Environ Sustain 1(1):2

Furukawa PDO, Cunha ICKO, Da Luz M, Pedreira G, Marck PB (2016) Environmental sustainability in medication processes performed in hospital nursing care. ACTA Paulista De Enfermagem 29(3):316–324. https://doi.org/10.1590/1982019420160043

Panizzolo R, Garengo P, Sharma MK, Gore A (2012) Lean manufacturing in developing countries: evidence from Indian SMEs. Prod Plan & Control: Manag Oper 23(10–11):769–788. https://doi.org/10.1080/09537287.2011.642155

Paranitharan, KP, Jeyathilagar D (2017) An empirical validation of integrated manufacturing business excellence model. Int J Adv Manuf Technol 92(5–8). Springer, 2569–2591

Paranitharan KP, Ramesh Babu T (2014) A comprehensive review of integrated manufacturing practice in global context-manufacturing management perspective. Int Bus Manag 8(6):369–377. https://doi.org/10.3923/ibm.2014.369.377

Pepper, Matthew PJ, Trevor A Spedding (2010) The evolution of lean six sigma. Int J Qual & Reliab Manag. Emerald Group Publishing Limited

Ruben, Ben, R, Vinodh, S, Asokan, P (2018) Lean six sigma with environmental focus: review and framework. Int J Adv Manuf Technol 94(9). Springer, 4023–4037

Rybkowski, Zofia K, Xun Zhou, Sarel Lavy, Jose Fernández-Solís (2012) Investigation into the nature of productivity gains observed during the airplane game lean simulation. Lean Constr J

Sajan MP, Shalij PR, Ramesh A, Biju AP (2017) Lean manufacturing practices in Indian manufacturing SMEs and their effect on sustainability performance. J Manuf Technol Manag 28(6):772–793. https://doi.org/10.1108/JMTM-12-2016-0188

Sakthi Nagaraj, T, R Jeyapaul (2020) An empirical investigation on association between human factors, ergonomics and lean manufacturing. Production Planning & Control. Taylor & Francis, 1–15

Shah R, Ward PT (2007) Defining and developing measures of lean production. J Oper Manag 25(4):785–805

Snee, Ronald D (2010) Lean Six Sigma-Getting Better All the time. Int J Lean Six Sigma. Emerald Group Publishing Limited

Sony, Michael, Subhash Naik (2020) Green lean six sigma implementation framework: a case of reducing graphite and dust pollution. Int J Sustain Eng 13(3). Taylor & Francis, 184–193. https://doi.org/10.1080/19397038.2019.1695015

Sony, Michael, Subhash Naik, Jiju Antony (2020) Lean six sigma and social performance: a review and synthesis of current evidence. Qual Manag J 27(1). Taylor & Francis, 21–36

Thomas, AJ, Hefin Rowlands, Paul Byard, Rhys Rowland-Jones (2008) Lean six sigma: an integrated strategy for manufacturing sustainability. Int J Six Sigma Compet Advant 4(4). Inderscience Publishers, 333–354

Thomas AJ, Ringwald K, Parfitt S, Davies A, John E (2014) An empirical analysis of Lean Six Sigma implementation in SMEs – a migratory perspective. Int J Qual & Reliab Manag 31(8):888–905. https://doi.org/10.1108/IJQRM-04-2013-0070

Touriki, Fatima Ezahra, Imane Benkhati, Sachin S Kamble, Amine Belhadi (2021) An integrated smart, green, resilient, and lean manufacturing framework: a literature review and future research directions. J Clean Prod. Elsevier, 128691

Wahab, Amelia Natasya Abdul, Muriati Mukhtar, Riza Sulaiman (2013) A Conceptual model of lean manufacturing dimensions. Procedia Technol 11. Elsevier, 1292–1298

WCED, SPECIAL WORKING SESSION (1987) World commission on environment and development. Our Common Futur 17(1):1–91

Wheat, Barbara, Chuck Mills, Mike Carnell (2003) Leaning into six sigma: a parable of the journey to six sigma and a lean enterprise. McGraw Hill Professional

White, RE, Pearson, JN, Wilson, JR (1999) JIT manufacturing: a survey of implementations in small and large U.S. manufacturers. Manag Sci, 45(1):1–15. https://doi.org/10.1287/mnsc.45.1.1

Wilkinson, Adrian (1998) Managing with Total Quality Management: theory and practice. Macmillan International Higher Education

Womack, J, Jones, D, Roos, D (1990) The machine that changed the world: the story of lean production. HarperCollins

Worley JM, Doolen TL (2015) Organizational structure, employee problem solving, and lean implementation. Int J Lean Six Sigma 6(1):39–58. https://doi.org/10.1108/IJLSS-12-2013-0058

Zhou B (2012) Lean principles, practices, and impacts: a study on small and medium-sized enterprises (SMEs). Ann Oper Res 1–18. https://doi.org/10.1007/s10479-012-1177-3

Zu, Xingxing, Lawrence D Fredendall, Thomas J Douglas (2008) The evolving theory of Quality Management: the role of six sigma. J Oper Manag 26(5). Elsevier, 630–650

Chapter 5
Beyond the Hype: Smart Manufacturing and Sustainable Excellence for SMEs

Objectives
The current chapter aims to:

1. Provide a brief description of smart manufacturing and its main technologies.
2. Highlight the positive impact of SM on sustainable business excellence for SMEs

5.1 Introduction

The concept of business excellence has grown worldwide as a new trend that takes TQM implementation frameworks and quality award schemes to the next level. However, the business excellence trend has primarily focused on enterprise longevity and financial security, leaving out the environment and societal dimensions (Edgeman and Eskildsen 2014). Nevertheless, in the last decades, organizations and businesses have been confronting a variety of mega forces and pressures to transform their business excellence approach into a sustainable approach that emphasizes on the economic, social, environmental and even ethical impact (Al-Qutop and Harrim 2014).

Small and medium enterprises (SMEs) on the other hand, despite the significant environmental impact that have and their contribution in the social welfare through the creation of employment opportunities (Sternad, Krenn, and Schmid 2019), they are still creeping behind larger corporations in the adoption of sustainable business excellences strategies. Past literature strongly argues that SMEs can also benefit from the sustainable business excellences approaches once they acknowledge certain limitations and challenges (Belhadi, Touriki, and El Fezazi 2018a).

In this special context, SMEs are called to deal with competitive pressures, act on sustainable dimensions, optimize the manufacturing process, eliminate waste and achieve continuous improvement (Cherrafi et al. 2016). In this vein, smart manufacturing (SM) emerges as a key element in order to respond the above challenges.

F. E. Touriki et al., *Sustainable Excellence in Small and Medium Sized Enterprises*, Industrial Ecology, https://doi.org/10.1007/978-981-19-0371-7_5

The numerous technologies provided in the SM environment will guide SMEs to consider new ways of sustaining their operations while at the same time preserving the environmental and social integrity (Touriki et al. 2021).

Therefore, in this paper, we try to take an overview of the smart sustainable business excellence approaches, benefits and opportunities for SMEs. We also highlight the characteristics that SMEs should acknowledge in order to sustainably excel. Finally, we showcase some key applications that would encourage SMEs to take the road of smart and sustainable business excellence.

5.2 Smart Manufacturing

There is no generally accepted definition of SM (Kusiak 2018). The National Institute of Standards and Technology (NIST) defines SM as a "fully integrated, collaborative manufacturing system that respond in real time to meet changing demands and conditions in the factory, in the supply network and in customer needs". SM aims to translate the captured data across the manufacturing process, from the product life cycle, machines and equipment, into an intelligence for governing manufacturing operations (Tao et al. 2018), it also, aspires prompt responses to complicated and diversified situations in real time (Kang et al. 2016). Similarly, Frank, Dalenogare, and Ayala (2019) considers SM as an "adaptable system where flexible lines adjust automatically production processes for multiple types of products and changing conditions".

Several technologies have emerged to be used in the smart manufacturing process (Dalenogare et al., 2018):

- The internet of things (IoT)

The technologies associated with this concept make it possible to collect data from machines at the heart of the plant or decentralized and feed them back into the information system. The goal of smart manufacturing is to offer better, more efficient and smarter production. To do this, it is necessary to have at all times the general state of all production systems in order to react intelligently to the various events that may arise. The IoT also works the other way around and allows orders or decisions to be sent back directly to machines.

- Cloud Computing

The cloud is made up of many technologies that make it possible to harness the computing and storage power of remote computer servers. So instead of having local IT infrastructures, the cloud makes it possible to outsource all or part of it. Often, rationalizing costs and billing per consumer are much more beneficial than owning and maintaining your equipment in-house.

- Big Data

Big data plays a central role in processing the enormous volume of information that today makes up an industry's information system. Indeed, IoT, optimization algorithms, ERP today make up a gargantuan volume of data. Conventional data storage technologies are now overtaken by these volumes. This is why suitable technologies have been created to store such volumes. There are several reasons for these growing amounts of data, but the development of the Internet and data exchange infrastructures are certainly at the origin of the creation of this need in companies. There are therefore now solutions that meet this need. However, this is of no use if there is no system capable of processing, analyzing and extracting this data.

- Blockchain

While the volume of data and the number of flows fed back by production chains, machines and various sensors or IoTs keep increasing, industrial applications require a high degree of trust and protection. Indeed, proof of the existence of the data source (sensor, machine, etc.) or of the consistency and veracity of the data collected becomes very important. How can we be sure that the data we receive from a sensor, for example, is real, unmodified in its path and coming from a single point of contact, verified, valid and expected?It is in this context that blockchain brings real value to data. Blockchain is a "cryptography-based computer protocol" that secures data transferred digitally from a single source of data to a single recipient. The blockchain brings confidence in the veracity of the source and the data through several aspects: "Everyone" (hear all "subscribers" of the blockchain) has a complete log of transactions and can check that the transfer has been madeNo one can refute the legitimacy of the transfer.

- Artificial Intelligence

Artificial intelligence (AI) is "the set of theories and techniques implemented to produce machines capable of simulating human intelligence". It therefore encompasses a set of concepts and technologies, more than an autonomous constituted discipline. Often classified in the group of mathematics and cognitive sciences, it uses computational neurobiology (particularly neural networks) and mathematical logic (part of mathematics and philosophy). It uses problem solving methods with high logical or algorithmic complexity. Artificial intelligence will no longer only be an actor in analysis but also becomes an actor in automated decision-making.

- Virtual Reality and Augmented Reality

Virtual reality (Virtual Reality—VR) and augmented reality (Augmented Reality—AR) bring together the digital, virtual and real world. Augmented reality is a technique that makes it possible to view objects or physical environments live in an improved manner using a digital medium (glasses, tablets, smartphones, etc.). The displayed video stream transcribes the physical world with additional information. This technology can play many roles in Industry 4.0 and bring real added value up the production chain.

- Robotics

Beyond automation, robots are increasingly agile, autonomous and capable of "intelligent" reactions and have a central place in certain industrial innovations. Integrated and connected to the entire information system, robots bring many benefits to the industry and are increasingly integrated into the "delicate" or "repetitive" procedures of the processes.

- 3D printing

3D printing, also called additive manufacturing, refers to a set of processes intended to produce objects by adding layers of materials corresponding to successive sections of a 3D model. Plastics and metal alloys are the most common materials used for 3D printing, but it can work on just about anything from concrete to living tissue.

According to Kusiak (2018), SM incorporates the available resources that the company currently have with the future generation of sensors, communication technology, computing platforms and data analytics. SM accordingly uses the concepts of cyber physical systems, the internet of things, cloud computing, artificial intelligence, data science, robotics, additive manufacturing, system integration, simulation and augmented reality (Kang et al. 2016; Kusiak 2018; Frank, Dalenogare, and Ayala 2019; Dalenogare et al. 2018). These technologies allow to improve quality, productivity, flexibility and high product customization at a large and sustained way (de Sousa Jabbour et al. 2018; Frank, Dalenogare, and Ayala 2019). Furthermore, they permit a reduction of resources utilization, delivery times and information distortion (Ivanov et al. 2014). Into the bargain, they enable companies to analyze a large amount of data which can highly facilitate the decision-making process (Porter and Heppelmann 2014). Moreover, SM technologies can monitor and control equipment, machines and products thorough a cycle of feedback while collecting a huge amount of data, which can help correcting and preventing any potential troubleshooting (Kusiak 2019).

Despite the litany of SM benefits, there are still some potential drawbacks that call for scholars and practitioners' attention. Examples include lack of understanding, high cost, potential energy consumption issues and significant uncertainty and varying performance results across different applications or contexts (Bai et al. 2020).

5.3 Sustainable Excellence

5.3.1 Business Excellence

Earlier definition by the European Foundation for Quality Management suggests that excellence is managing organizations using outstanding practices in order to achieve the expected outcomes (Tarí and Molina-Azorín 2010). Accordingly, an excellent firm strives to reach superior, sustainable and comprehensive achievements that go hand in hand with the stakeholders' requirements (Mangla et al. 2020).

Excellent firms balance internal and external requirements for performance, they have better leadership and governance, they also achieve higher financial results and hold healthy and sustainable relationships with their customers and suppliers; they meet the interest of their employees, their stakeholders and the community (Jankalová and Jankal 2020).

In today's modern manufacturing, lean manufacturing, lean six sigma, quality management approaches and the technologies of the industry 4.0 (I4.0) are frequently related to the business excellence programs (Kamble, Gunasekaran, and Sharma 2020; Belhadi et al. 2020; Cherrafi et al. 2017). In this chapter we will focus on Lean, LSS as the main strategies for business excellence in addition to the impact of SM to achieve the latter.

5.3.2 Sustainability

The World Commission on Environment and Development's 1987 Brundtland report defines sustainability as a "development that meets the needs of the present without compromising the ability of future generations to meet their own needs" (WCED 1987). It is often referred to as the capacity of the earth's biosphere and human civilization to coexist. According to the U.S. National Environmental Policy Act of 1969 (NEPA), the goal of sustainability is to, "create and maintain conditions, under which humans and nature can exist in productive harmony, that permit fulfilling the social, economic, and other requirements of present and future generations."

In the context of manufacturing, companies are called to achieve maximum profits, endure market changes and volatility, reduce the harmful impacts of manufacturing operations on the environment and guarantee the safety, well-being and welfare of their employees (Elkington 1994) Under this conception, sustainability could be defining through three main pillars, often referred to as the triple bottom line, where all three: economic aspect, environmental aspect, and social aspect need to be balanced (Ruddy and Hilty 2008).

It might seem that the economic pillar might surpass the other pillars, however according to Morelli (2011) all three pillars are interdependent, ultimately, they will all complement each other. Therefore, manufacturers need to equally focus on all pillars in order to achieve sustainability.

5.3.3 Sustainable Business Excellence

Recently, being profitable is not enough to sustain in a competitive environment. Firms are enduring several mega forces. In the present scenario, they are called to comply with customers, regulators and stakeholders requirements and calls for sustainability. Climate change, poverty, scare resources, pollution, unemployment and inequity are all factors that put more pressures on manufacturers to improve their environmental and social performance in a responsible manner (Garza-Reyes 2015). Consequently, sustainability is no longer a choice or a just a "nice to have", rather it is a necessity and a "must have", therefore, companies need to come up with sustainable management strategies in order to tackles all these challenges (Kamble, Gunasekaran, and Sharma 2018).

Lean manufacturing and six sigma as both programs of business excellence have been reported to have positive effects on sustainability (Belhadi et al. 2018b; Kamble, Gunasekaran, and Sharma 2020; Belhadi et al. 2020; Cherrafi et al. 2017). Therefore, firms can move from operational business excellence to sustainable business excellence.

LM has been defied in different ways; however, the main goal of lean is to eliminate waste while delivering high quality products and services at the lowest cost and time (Bhasin and Burcher 2006).

Across the literature, the economic pillar is the most discussed one. Shah and Ward (2007) elaborated those lean practices can lead to significant improvements in operational performance up to 20%. In addition, lean can maximize and sustain business profitability, meet customer needs and satisfaction, and reduce operational costs of sustainability (Büyüközkan, Kayakutlu, and Karakadılar 2015, Henao, Sarache, and Gómez 2019, Belhadi, Touriki, and El Fezazi 2018a). Literature has also highlighted the advantages of lean on the environmental level. Authors like Fatemi and Franchetti (2016) stated the existence of improvements in cost and environmental criteria through lean applications. On the social level, despite the scarcity of literature discussing the impact of lean on the social dimension of sustainability, most agree that lean practices can improve workers satisfaction and comfort. Moreover, lean can enhance the working conditions including health and safety and ergonomic. It can also improve teamwork and connect employees (Henao, Sarache, and Gómez 2019; Sony and Naik 2020).

First introduced by Bill Smith at Motorola Corporation, in the 1980s, six sigma evolved to be a data driven approach that aims to reduce variation, errors and defects (Pepper and Spedding 2010). The reported benefits of six sigma vary from increasing the overall performance, improving customer satisfaction and prevent products defects (Snee 2010). Benefits also include the environmental aspect; a study by (Furukawa et al. (2016) indicated a reduction in packaging and plastic bags. Besides, socially, Sony et al. 2020 determined that six sigma projects have positive influence on decent work conditions employee's well-being and satisfaction.

5.4 Small and Medium Enterprises

SMEs are a category of businesses whose number of employees, revenues or asset fall below a certain threshold (Loecher 2000). Nevertheless, there is no common definition of SMEs since size threshold differ between countries and industries (Baba et al. 2006). However the European Union (EU) provides a clear delineation of a small sized enterprise as a firm with fewer than 50 employees and a medium sized enterprise as one with less than 250 employees. Micro enterprises employ a range of 10 employees. Table 5.1 provides a summary of SMEs definition by country.

SMEs play a crucial role in driving economic growth, innovation and competitiveness (Antoldi, Cerrato, and Depperu 2012). They also employ a large percentage of the workforce and generate significant added value especially in the developing countries (Scoreboard 2015).

The first apparent characteristic of SMEs is their low size. This latter is directly related to low complexity, less interfaces between departments and flexible and flat structure (Ghobadian and Gallear 1996). Next, we find a high typological diversity since SMEs are operational everywhere within each country (Nicolescu 2009). Into the bargain, SMEs are human intensive, since a small number of employees is involved, therefore human interactions are more frequent and direct, hence, they are more conductive to change and have fluent communication. In addition, SMEs are flexible in structure; they have high capacity of adjustment to the needs and requirements since they have less complexity to deal with (Baba et al. 2006).

On the other side, most SMEs have low level of formalization, this is manifested in lack of documentary system, procedures and job descriptions. This can hinder the implementation of new projects such as lean or six sigma that highly depend on an effective documentation system. Moreover, many SMEs frequently use an authoritarian management approach which can jeopardize the implication and motivation of employees in the lean six sigma projects, hence, SMEs need to adopt the participative managerial approach that supports the lean thinking (Bartlett and Bukvič 2001). Finally, the decision-making process in SMEs is highly centralized. According to Nicolescu (2009), an SME is mainly the result of the owner or the entrepreneur efforts and capability, therefore most decisions are made by one single person which can block or delay the LSS projects.

Increased competition, rapidly changing market, technological changes and other factors are among the mega forces that deter the growth of SMEs. Many SMEs face great difficulties in accessing finance, since financial institutions are less disposed to allocate their resources to SMEs (Yoshino and Taghizadeh Hesary 2016). SMEs are usually regarded as riskier enterprises compared to larger companies that have lower risk of default and clear financial statements (Stiglitz and Weiss 1981). Consequently, inadequate supply of financial capital hinders the survival and growth of SMEs (Muriithi 2017). As a result, SMEs have little spare resources in terms of employment, therefore each person has a key duty, most often numerous. Hence, projects like lean and six sigma can be a struggle to deal with, since dedicating specific employees a certain project will leave them less time for their actual work.

Another challenge that faces SMEs is poor management. Many SMEs are owned by people who usually lack appropriate training and experience to operate their businesses. In addition, they are driven by short term performance and gain which can hinder the continuous improvement projects unlike large organizations who have medium and long-term vision (Muriithi 2017). Effective training and education can empower SMEs managers to acquire soft skills that can enable them to succeed and build good business practices (King and McGrath 2002).

Into the bargain, the era of new information technology has developed rapidly. However, SMEs have been unable to sufficiently make use of such opportunities (Yoshino and Taghizadeh Hesary 2016). Besides, the cost of these technologies and the highly-skilled personnel that comes with it, represent another barrier that stops SMEs from fully utilizing the new technologies of the I4.0 (Niaki and Nonino 2017). Furthermore, negative values in SMEs, relentless beliefs, and resistance to change or adoption of these technologies are owing to the fear of complexity in the use of the latest technology and the lack of confidence among the staff and managers, which can delay future projects of business excellence supported by I4.0 (Haseeb et al. 2019).

5.5 Smart Manufacturing and Sustainable Excellence for SMEs

5.5.1 Smart Sustainable Excellence for SMEs

Throughout the literature, there is a mutual agreement that the era of digital technologies holds great potential in promoting the projects of business excellence for SMEs. Most scholars and practitioners identified a strong connection between I4.0 and lean paradigms. Studies consider lean to be a prerequisite of SM; this later also stands as a tool to overcome the barriers and limitations of lean (Kolberg and Zühlke 2015). According to Moeuf et al. (2020) I4.0 technologies can improve industrial performance, flexibility, productivity, delivery time cost and quality which is perfectly in line with lean manufacturing objectives. Sanders, Elangeswaran, and Wulfsberg (2016) also suggests that the solutions provided by I4.0 will boost SMEs performance and enable them to become lean without "striving-for-lean" efforts.

Furthermore, the power of advanced analytics brought by SM will enhance the six sigma approaches, since SS is a data driven methodology. Data collection, data treatment and processing, root causes analysis and action plans elaboration are now easier and less time consuming thanks to I4.0. As a result, managers can make effective decisions for quality problems by benefiting from advanced techniques such as big data analytics and process mining in addition to traditional techniques (Dogan and Gurcan 2018).

Moreover, I4.0 can help both SMEs and society to move towards sustainable development (de Sousa Jabbour et al. 2018) through environmental protection integration, control initiatives, process safety, resources efficiency approaches, employee and community welfare and smarter and flexible processes (Luthra and Mangla 2018). Consequently, SM transforms the production system and the supply chain into a smart environment based on CPS that is flexible profitable and eco-friendly.

According to (Kamble et al. 2018), manufacturing organizations must surpass pure profit maximization, and address broader societal and environmental expectations. In this regard, digital transformation must become a "legitimacy goal", where the technologies of SM will help meet these expectations.

In table 5.1, we present the potential relationships between SM technologies and the Sustainable Development Goals elaborated by the United Nations as presented by Bai et al., 2020

However, despite the great role of SM in improving sustainable operations, there is still uncertainty and confusion since SM applications can have a different and contradictory outcome in some cases. According to Bai et al. 2020, the application of SM and digitization could improve equality and economic business items but they require additional energy requirements, which can result in resource depletion or environmentally damaging emissions.

In this vein, authors proposed a multi-contextual decision-making method for evaluating the impact of I4.0 T on the three sustainability dimensions, which can be applied for evaluating sustainability priorities and which technologies to adopt.

In nutshell, the applications of SM technologies for sustainable business excellence seem to correlate positively with today's business requirements, therefore both practitioners and scholars need to consider investigating further the impact of SM on SBE and how SMEs could benefit from this integration.

5.5.2 Case Studies

5.5.2.1 Smart Machines

In a smart factory, machines are smart and interconnected. In the case of a breakdown, the machines send error alarm and notifications to the maintenance engineers where they check the error for troubleshooting and mitigate the disruption (Lucke, Constantinescu, and Westkämper 2008). Machines are designed to be self-aware; they check their own status and use data for preventive action plans. As a result, potential maintenance issues are avoided and the employee's safety is enhanced (Lee, Kao, and Yang 2014). In addition, smart machines are based on RFID solutions which can be used for the detection of tagged Kanban cards in real-time (Mrugalska and Wyrwicka 2017).

Enterprise resource planning systems are another application of the era of I4.0 in enhancing sustainable excellence projects. With their comprehensive frameworks for predictive maintenance, they integrate data coming from machines, ERP, sensors and predictive algorithms (Haddara and Elragal 2015).

Table 5.1 The United Nations Sustainable Development Goals and Industry 4.0 technologies *Source* Bai et al. 2020

Goal	Explanation	Relationship with Industry 4.0 technologies
End poverty	End poverty in all its forms everywhere	Industry 4.0 technologies can bring access to information, education, health care and greater economic opportunity that provide more basic resources and services to the poor people and bring them out of poverty. Industry 4.0 technologies also can alleviate the unexpected economic losses during disasters
End hunger	End hunger, achieve food security and improved nutrition and promote sustainable agriculture	Industry 4.0 technologies can promote sustainable agriculture and fair distribution systems to make sure that nobody will ever suffer from hunger again. Industry 4.0 technologies also help to achieve food security and improved nutrition
Good health and well-being	Ensure healthy lives and promote well-being for all at all ages	Industry 4.0 technologies are enabling the digitalization of healthcare, it promotes healthy lifestyles, preventive measures and modern, efficient healthcare services for everyone
Quality education	Ensure inclusive and equitable quality education and promote lifelong learning opportunities for all	Industry 4.0 technologies can help enjoy affordable, equitable and quality primary and secondary education
Gender equality	Achieve gender equality and empower all women and girls	Industry 4.0 technologies will provide the equal opportunity to succeed in every function and at every level for both men and women
Clean water and sanitation	Ensure availability and sustainable management of water and sanitation for all	Industry 4.0 technologies can help provide affordable equipment and education in hygiene practices that access to clean drinking water and sanitation

(continued)

Table 5.1 (continued)

Goal	Explanation	Relationship with Industry 4.0 technologies
Affordable and clean energy	Ensure access to affordable, reliable, sustainable and modern energy for all	Industry 4.0 technologies will bring sustainable energy efficiency to enable both increased quality and cost savings to users
Decent work and economic growth	Promote inclusive and sustainable economic growth, employment and decent work	Industry 4.0 technologies have significant direct and indirect economic impact while not harming the environment. Economic growth will create decent and fulfilling jobs
Industry, innovation and infrastructure	Build resilient infrastructure, promote sustainable industrialization and foster innovation	Industry 4.0 technologies, be as new industries and information and communication technologies, are the important ways to promoting sustainable industries, investing in scientific research and innovation, upgrading infrastructure
Reduced inequalities	Reduce inequality within and among countries	Industry 4.0 technologies can help to connect the unconnected and helps to bridge the digital divide to reduce inequality within and among countries or people
Sustainable cities and communities	Make cities inclusive, safe, resilient and sustainable	Industry 4.0 technologies can help build modern, sustainable, intelligent, public order, safety and security cities with green and culturally inspiring living conditions
Responsible consumption and production	Ensure sustainable consumption and production patterns	Industry 4.0 technologies are already being used to bring greater efficiency across industries, greater transparency in supply chains, and improved production and consumption patterns

(continued)

Table 5.1 (continued)

Goal	Explanation	Relationship with Industry 4.0 technologies
Climate action	Take urgent actions to combat climate change and its impacts	Industry 4.0 technologies can lead to a reduction of CO2-emissions by data-centered and traceable carbon footprint
Life Below Water	Conserve and sustainably use the oceans, seas and marine resource	Industry 4.0 technologies can contribute to more sustainable use of the oceans resources
Life on land	Sustainably manage forests, combat desertification, halt and reverse land degradation, halt biodiversity loss	Industry 4.0 technologies are key to management of forests and biodiversity
Peace, justice and strong institutions	Promote just, peaceful and inclusive societies	Industry 4.0 technologies can ensure and promote a responsible firm or supply chain, human rights and freedom of expression
Partnerships for the goals	Revitalize the global partnership for sustainable development	Industry 4.0 technologies can help collaborate with a broad range of stakeholders in order to drive sustainable development forward

5.5.2.2 Smart Products

One of the objectives of the sustainable excellence is to guarantee products quality. With the six-sigma approach, processes are always under control. In the era of SM, the tracking of process variations for quality defects detection is enhanced. In this context, products are smart, therefore, the operation to performed on a product is already loaded into that product. This latter is transmitted to the machine and then visualized (Sanders, Elangeswaran, and Wulfsberg 2016).

Smart products will also collect and use data collected from sensors for actions plans. According to Mrugalska and Wyrwicka (2017), smart products are "context-aware, adaptive, self-organized, and proactive and able to support the whole life-cycle". This can enhance the continuous improvement process, visualization, and Kanban initiatives.

5.5.2.3 Smart Production

In sustainable excellence, the overall production flow should be adaptable to the pull system. Thanks to the information and communication technologies, an e-kanban system can be of a great use in supporting automatic replenishment monitoring and schedule tracking. It identifies the missing or empty bins with smart sensors, transfers data in real time and monitors the status, number and location of material batches (Kouri, Salmimaa, and Vilpola 2008).

5.5.2.4 Smart Relationships

In the context of smart sustainable excellence, the relationship with suppliers and customers is smart. I4.0 provides the necessary tools to improve suppliers and customer's feedback; collaborative manufacturing in the context of SM adequately serves these purposes, especially for SMEs with limited resources.

This collaboration approach can also mitigate risks in case of disruptions which is in line of the objectives of sustainability (Mrugalska and Wyrwicka 2017). Through cloud computing, the traditional ways for communication between suppliers, SMEs and customers are renovated (Schmidt et al. 2015). Thus, more collaborative, synchronized, affordable and improved mechanisms are provided to enable SMEs to maintain effective relationships with both suppliers and customers.

5.6 Conclusion

In the present context, sustainability has become a business imperative, therefore companies are asked to transform their traditional ways of working and move from operational business excellence to sustainable business excellence. In this paper we highlight the role of SM in achieving sustainable business excellence for SMEs. We also showcase some key applications where SM has proven to be beneficial for real manufacturing shop floor for the achievement of both sustainable and operational business excellence. We hope that this paper will encourage SMEs into considering the adoption of I4.0 for sustainable manufacturing.

References

Al-Qutop, Mohi-Adden Yahya, Hussein Harrim (2014) Spiritual organization: a path to organizational sustainable excellence. *Int Bus Res* 7(8). Canadian Center of Science and Education, 168

Antoldi, Fabio, Daniele Cerrato, Donatella Depperu (2012) Export consortia in developing countries: successful management of cooperation among SMEs. Springer Science & Business Media

Baba, Md Deros, Sha'ri Mohd Yusof, Azhari, Md Salleh (2006) A benchmarking implementation framework for automotive manufacturing SMEs. *Benchmarking: Int J* 13(4). Emerald Group Publishing Limited, 396–430

Bai, Chunguang, Patrick Dallasega, Guido Orzes, Joseph Sarkis (2020) Industry 4.0 technologies assessment: a sustainability perspective. *Int J Prod Econ* 229. Elsevier, 107776

Bartlett, Will, Vladimir Bukvič (2001) Barriers to SME growth in Slovenia. *MOST: Econ Policy TransitAl Econ* 11(2). Springer, 177–195

Belhadi A, Kamble SS, Zkik K, Cherrafi A, Touriki FE (2020) The integrated effect of big data analytics, lean six sigma and green manufacturing on the environmental performance of manufacturing companies: the case of North Africa. J Clean Prod 252. https://doi.org/10.1016/j.jclepro.2019.119903

Belhadi, Amine, Yusof Bin Mohd Sha'ri, Fatima Ezahra Touriki, Said El Fezazi (2018a) Lean Production in SMEs: Literature Review and Reflection on Future Challenges. *Journal of Industrial and Production Engineering* 35(6). Taylor & Francis, 368–382

Belhadi, Amine, Yusof Bin Mohd Sha'ri, Fatima Ezahra Touriki, Said El Fezazi (2018b) Benefits of adopting lean production on green performance of SMEs: a case study. *Prod Plan & Control* 29(11). Taylor & Francis, 873–894

Bhasin, Sanjay, Peter Burcher (2006) Lean viewed as a philosophy. *J Manuf Technol Manag.* Emerald Group Publishing Limited

Büyüközkan, Gülçin, Gülgün Kayakutlu, İbrahim S Karakadılar (2015) Assessment of lean manufacturing effect on business performance using bayesian belief networks. *Expert Syst Appl* 42(19). Elsevier, 6539–6351

Cherrafi, Anass, Said Elfezazi, Andrea Chiarini, Ahmed Mokhlis, Khalid Benhida (2016) The integration of lean manufacturing, six sigma and sustainability: a literature review and future research directions for developing a specific model. *J Clean Prod* 139. Elsevier, 828–846

Cherrafi, Anass, Said Elfezazi, Andrea Chiarini, Ahmed Mokhlis, Khalid Benhida (2017) Exploring critical success factors for implementing green lean six sigma. *Int Manuf Strat Time Gt Flux*, 183–195. Springer

Dalenogare, Lucas Santos, Guilherme Brittes Benitez, Néstor Fabián Ayala, Alejandro Germán Frank (2018) The expected contribution of industry 4.0 technologies for industrial performance. *Int J Prod Econ* 204. Elsevier, 383–394

Dogan, Onur, Omer Faruk Gurcan (2018) Data perspective of lean six sigma in industry 4.0 era: a guide to improve quality. *Proc Int Conf Ind Eng Oper Manag* 2018 (JUL), 943–953

Edgeman, Rick, Jacob Eskildsen (2014) Modeling and assessing sustainable enterprise excellence. *Bus Strat Environ* 23(3). Wiley Online Library, 173–187

Elkington, John (1994) Towards the sustainable corporation: win-win-win business strategies for sustainable development. *Calif Manag Rev* 36(2). SAGE Publications Sage CA: Los Angeles, CA, 90–100

Fatemi, Shila, Matthew J Franchetti (2016) An application of sustainable lean and green strategy with a six sigma approach on a manufacturing system. *Int J Six Sigma Compet Advant* 10(1). Inderscience Publishers (IEL), 62–75

Frank, Alejandro Germán, Lucas Santos Dalenogare, Néstor Fabián Ayala (2019) Industry 4.0 technologies: implementation patterns in manufacturing companies. *Int J Prod Econ* 210. Elsevier, 15–26

Garza-Reyes, Jose Arturo (2015) Lean and green–a systematic review of the state of the art literature. *J Clean Prod* 102. Elsevier, 18–29

Ghobadian, Abby, David N Gallear (1996) Total Quality Management in SMEs. *Omega* 24(1). Elsevier, 83–106

Haddara, Moutaz, Ahmed Elragal (2015) The readiness of ERP systems for the factory of the future. *Procedia Comput Sci* 64. Elsevier Masson SAS, 721–728. https://doi.org/10.1016/j.procs.2015.08.598

Haseeb, Muhammad, Hafezali Iqbal Hussain, Sebastian Kot, Armenia Androniceanu, Kittisak Jermsittiparsert (2019) Role of social and technological challenges in achieving a sustainable competitive advantage and sustainable business performance. *Sustainability* 11(14). Multidisciplinary Digital Publishing Institute, 3811

Henao, Rafael, William Sarache, Iván Gómez (2019) Lean manufacturing and sustainable performance: trends and future challenges. *J Clean Prod* 208. Elsevier, 99–116

Ivanov, Dmitry, Boris Sokolov, Alexandre Dolgui (2014) The ripple effect in supply chains: trade-off 'efficiency-flexibility-resilience' in disruption management. *Int J Prod Res* 52(7). Taylor & Francis, 2154–2172

Jankalová, Miriam, Radoslav Jankal (2020) How to characterize business excellence and determine the relation between business excellence and sustainability. *Sustainability* 12(15). Multidisciplinary Digital Publishing Institute, 6198

Kamble, Sachin S, Angappa Gunasekaran, Rohit Sharma (2018) Analysis of the driving and dependence power of barriers to adopt industry 4.0 in Indian manufacturing industry. *Comput Ind* 101. Elsevier, 107–119

Kamble, Sachin S, Angappa Gunasekaran, Rohit Sharma (2020) Modeling the blockchain enabled traceability in agriculture supply chain. *Int J Inf Manag* 52. Elsevier, 101967

Kang, Hyoung Seok, Ju Yeon Lee, SangSu Choi, Hyun Kim, Jun Hee Park, Ji Yeon Son, Bo Hyun Kim, Sang Do Noh (2016) Smart manufacturing: past research, present findings, and future directions. *Int J Precis Eng Manuf-Green Technol* 3(1). Springer, 111–128

King, Kenneth, Simon McGrath (2002) Globalisation, enterprise and knowledge: education, training and development in Africa. Symposium Books Ltd

Kolberg D, Zühlke D (2015) Lean automation enabled by industry 4.0 technologies. IFAC-PapersOnLine 28(3):1870–1875. https://doi.org/10.1016/j.ifacol.2015.06.359

Kouri IA, Salmimaa TJ, Vilpola IH (2008) The principles and planning process of an electronic kanban system. *Nov Algorithms Tech Telecommun, Autom Ind Electron*, 99–104. Springer

Kusiak, Andrew (2018) Smart manufacturing. *Int J Prod Res* 56(1–2). Taylor & Francis, 508–517

Kusiak, Andrew (2019) Fundamentals of smart manufacturing: a multi-thread perspective. *Annu Rev Control* 47. Elsevier Ltd, 214–220. https://doi.org/10.1016/j.arcontrol.2019.02.001

Lee, Jay, Hung An Kao, Shanhu Yang (2014) Service innovation and smart analytics for industry 4.0 and big data environment. *Procedia CIRP* 16. Elsevier B.V., 3–8. https://doi.org/10.1016/j.procir.2014.02.001

Loecher, Ulrich (2000) Small and medium-sized enterprises–delimitation and the european definition in the area of industrial business. *Eur Bus Rev*. MCB UP Ltd

Lucke, Dominik, Carmen Constantinescu, Engelbert Westkämper (2008) Smart factory—a step towards the next generation of manufacturing. *Manuf Syst Technol New Front*, no. Sfb 627, 115–118.https://doi.org/10.1007/978-1-84800-267-8_23

Luthra, Sunil, Sachin Kumar Mangla (2018) Evaluating challenges to industry 4.0 initiatives for supply chain sustainability in emerging economies. *Process Saf Environ Prot* 117. Elsevier, 168–179

Mangla, Sachin Kumar, Simonov Kusi-Sarpong, Sunil Luthra, Chunguang Bai, Suresh Kumar Jakhar, Sharfuddin Ahmed Khan (2020) Operational excellence for improving sustainable supply chain performance. *Resour, Conserv, Recycl* 162. Elsevier, 105025

Moeuf, Alexandre, Samir Lamouri, Robert Pellerin, Simon Tamayo-Giraldo, Estefania Tobon-Valencia, Romain Eburdy (2020) Identification of critical success factors, risks and opportunities of industry 4.0 in SMEs. *Int J Prod Res* 58(5). Taylor & Francis, 1384–1400. https://doi.org/10.1080/00207543.2019.1636323

Morelli J (2011) Environmental sustainability: a definition for environmental professionals. J Environ Sustain 1(1):2

Mrugalska, Beata, Magdalena K. Wyrwicka (2017) Towards lean production in industry 4.0. *Procedia Eng* 182. The Author(s), 466–473. https://doi.org/10.1016/j.proeng.2017.03.135

Muriithi, Samuel M (2017) African small and medium enterprises (SMEs) contributions, challenges and solutions. Eur J Res Reflect Manag Sci

Niaki MK, Nonino F (2017) Impact of additive manufacturing on business competitiveness: a multiple case study. J Manuf Technol Manag 28(1):56–74. https://doi.org/10.1108/JMTM-01-2016-0001

Nicolescu, Ovidiu (2009) Main features of SMEs organisation system. *Rev Manag Comp Internațional* 10(3). EDITURA ASE, 405–413

Furukawa, Patricia de Oliveira, Cunha ICKO, Da Luz M, Pedreira G, Marck PB (2016) Environmental sustainability in medication processes performed in hospital nursing care. ACTA Paulista De Enfermagem 29(3):316–324. https://doi.org/10.1590/19820194201600043

Pepper, Matthew PJ, Trevor A Spedding (2010) The evolution of lean six sigma. *Int J Qual & Reliab Manag*. Emerald Group Publishing Limited

Porter ME, Heppelmann JE (2014) How smart, connected products are transforming competition. Harv Bus Rev 92(11):64–88

Ruddy, Thomas F, Lorenz M Hilty (2008) Impact assessment and policy learning in the european commission. *Environ Impact Assess Rev* 28(2–3). Elsevier, 90–105

Sanders A, Elangeswaran C, Wulfsberg J (2016) Industry 4.0 implies lean manufacturing: research activities in industry 4.0 function as enablers for lean manufacturing. J Ind Eng Manag 9(3):811–833. https://doi.org/10.3926/jiem.1940

Schmidt, Rainer, Michael Möhring, Ralf-Christian Härting, Christopher Reichstein, Pascal Neumaier, Philip Jozinović (2015) Industry 4.0-potentials for creating smart products: empirical research results. *Int Conf Bus Inf Syst*, 16–27. Springer

Scoreboard, OECD (2015) Financing SMEs and entrepreneurs

Shah, Rachna, Peter T Ward (2007) Defining and developing measures of lean production. *J Oper Manag* 25(4). Elsevier, 785–805

Snee, Ronald D (2010) Lean six sigma–getting better all the time. *Int J Lean Six Sigma*. Emerald Group Publishing Limited

Sony, Michael, Subhash Naik (2020) Green lean six sigma implementation framework: a case of reducing graphite and dust pollution. *Int J Sustain Eng* 13(3). Taylor & Francis, 184–193. https://doi.org/10.1080/19397038.2019.1695015

Sousa Jabbour, Ana Beatriz Lopes de, Charbel Jose Chiappetta Jabbour, Moacir Godinho Filho, David Roubaud (2018) Industry 4.0 and the circular economy: a proposed research agenda and original roadmap for sustainable operations. *Ann Oper Res* 270(1). Springer 273–286

Sternad, Dietmar, Melanie Krenn, Stefanie Schmid (2019) Business excellence for SMEs: motives, obstacles, and size-related adaptations. *Total Qual Manag & Bus Excel* 30(1–2). Taylor & Francis, 151–168

Stiglitz, Joseph E, Andrew Weiss (1981) Credit rationing in markets with imperfect information. *Am Econ Rev* 71(3). JSTOR, 393–410

Tao, Fei, Qinglin Qi, Ang Liu, Andrew Kusiak (2018) Data-driven smart manufacturing. *J Manuf Syst* 48 (July). The Society of Manufacturing Engineers, 157–169. https://doi.org/10.1016/j.jmsy.2018.01.006

Tarí JJ, Molina-Azorín JF (2010) Integration of quality management and environmental management systems: similarities and the role of the EFQM model. Emerald Group Publishing Limited, TQM J

Touriki, Fatima Ezahra, Imane Benkhati, Sachin S Kamble, Amine Belhadi (2021) An integrated smart, green, resilient, and lean manufacturing framework: a literature review and future research directions. *J Clean Prod*. Elsevier, 128691

WCED, SPECIAL WORKING SESSION (1987) World commission on environment and development. *Our Common Future* 17(1):1–91

Yoshino, Naoyuki, Farhad Taghizadeh Hesary (2016) Major challenges facing small and medium-sized enterprises in Asia and solutions for mitigating them. ADBI Working Paper 564

Chapter 6
The Role of Smart Manufacturing for the Integration of Lean, Six Sigma and Social Sustainability

Objectives

1. Highlight the role of smart manufacturing in fostering the relationship between lean, six sigma and social sustainability in SMEs.
2. Provide SMEs with a framework for the integration of smart manufacturing, lean six sigma for higher social sustainability in SMEs.

6.1 Introduction

The past literature alongside the most common practitioner-oriented conceptual frameworks of sustainability have historically been mainly focused on the economic and environmental aspects, however the social facet is still not well explored (Hutchins and Sutherland 2008). On the other hand, frameworks developing these aspects were developed and adapted for large companies, while SMEs are still lagging behind larger businesses in the adoption of sustainable frameworks especially with the strong evidence that these frameworks can benefit companies of all types and sizes in terms of improving their sustainable performance including economic, environmental and societal aspects. SMEs can conquer a competitive advantage, through social sustainability (SS), only if effort is done on the development of practices that are convenient to the specific SME and its profile, not through generic strategies and assumptions.

Managerial practices like lean manufacturing (LM) and lean six sigma (LSS) were found to improve the three pillars of sustainability including the social one (Sajan et al. 2017). However, implementing LM, LSS and SS in SMEs is faced by various barriers. SMEs are called to understand the benefits, opportunities and prevailing challenges that come with LM, LSS and SS in order to successfully implement and maintain both operational and social performance. Therefore, they need to implement advanced tools that can foster this implementation. The new era of smart manufacturing (SM), enabled by the novel technologies of industry 4.0 (I4.0), has been found to positively

F. E. Touriki et al., *Sustainable Excellence in Small and Medium Sized Enterprises*, Industrial Ecology, https://doi.org/10.1007/978-981-19-0371-7_6

impact the relationship between the three paradigms of LM, LSS and SS (Varela et al. 2019). Both organizations and societies can move towards sustainable development by the means of the digital transformation (de Sousa Jabbour et al. 2018).

The objective of this chapter is to elaborate on the role of SM in enabling LM, LSS and SS. Therefore, we propose a unique integrated framework for the integration of LM, LSS and SS supported by SM features along with the set of critical success factor for the success of this implementation.

6.2 Background

6.2.1 Social Sustainability

Unlike, the economic and environmental aspects of sustainability, the social facet is still not well defined (Hutchins and Sutherland 2008). Past literature shows that there is a clear debate on the definition of social sustainability (SS), its scope, objectives and indicators (Littig and Griessler 2005; Weingaertner and Moberg 2014). This disagreement has resulted in a wide spectrum of definitions and perspectives (Vallance et al. 2011). Hollander et al. (2016) defined SS as a positive trait of a society that promote long lasting conditions for human wellbeing. Vallance et al. (2011) presented SS as a holistic approach that deals with human and social capital management, acknowledges stakeholders and seeks to warrant sustainable benefits for the society, while taking into consideration the remaining Triple Bottom Line (TBL) components. A simpler definition was provided by Carroll (1991) that considered SS as simply "conducting business in a way that is consistent with morals and values of society but not necessary required by Law".

Aforesaid, the noted debate in the literature has arisen a variety of foundations and tools found that SS must rely on values of equity and democracy, while Koning (2002) claimed that SS needs to focus on culture equity and social justice. Sloan (2010) asserts that SS is based on preserving fair management practices across supply chains.

Social homogeneity, equitable income distribution, access to goods and services, socially responsible practices, engaging codes of conduct, third-party standards and supplier development programs are examples of tools enabling the successful implementation of SS (Suppliers n.d.).

In the quest for sustainability, the economic and environmental sustainability can be promoted simultaneously with social factors such as fairness, poverty, health, education, demography and employee engagement (Ajmal et al. 2018). Thus, progressing towards sustainable development requires achieving human wellbeing together with eco system integrity, not one at the cost of the other (Littig and Griessler 2005).

What is corporate social responsibility?

Even if it is not a new concept, the publication of the Iso 26000 standard in 2010, allowed to give an international definition, resulting from a consensus between several countries and therefore between several angles. Corporate Social Responsibility (CSR) is defined as follows: "an *organization's responsibility for the impacts of its decisions and activities on society and the environment, through transparent and ethical behavior that:*

- *contributes to Sustainable Development, including health and the welfare of society.*
- *considers the expectations of stakeholders.*
- *follows applicable law and consistent with international norms of behavior.*
- *and is integrated throughout* the *organization and implemented in its relations.*"

According to this definition, an organization is "Responsible" from the moment that its activities generate impacts on society and the environment. The authors (Aubrun et al. 2010) specify that there can be no societal responsibility if the organization's decisions and activities do not have an impact or consequences on society and the environment. Therefore, it is the existence of an impact that triggers responsibility. It is also interesting to consider the scope of this "Responsibility". The latter can admit two essential meanings: assuming a burden and/or being accountable (Cadet 2014). Indeed, this term covers two realities: it is both about assuming responsibility for one's decisions, activities and their impacts, and being accountable for them (Aubrun et al. 2010). Thus, the concept of CSR is based on the presumed existence of an implicit contract between a company and society, according to which the company has obligations towards society, which has the right to control it (Forman et al. 2007). With this reasoning, the company is evaluated not only according to its economic and financial results, but also through its social commitments put into practice through actions that promote the respect of fundamental human rights at work and the collective interests of the society in which it is located (Forman et al. 2007).

However, it is wise to also clarify the terms associated with the term CSR, should we use "societal" or "social" responsibility? According to (Yedder and Farhoud 2009), "social" concerns the conditions that make it possible to make the contributions and rewards of employees equitable; societal, on the other hand, concerns the relationship with the company's local communities (Capron and Quairel-Lanoizelée 2004).

According to the standard definition of CSR, it is agreed upon that this approach encompasses the impacts of the company's activities in two dimensions: social and environmental. In the following, only the social aspect is taken into consideration. With the spread of the notion of sustainable excellence, the social dimension remains poorly covered in terms of methods and guides, allowing companies to affirm their commitment to society (Forman et al. 2007). To fill this gap, the ISO 26000 standard proposes two fundamental practices to its followers to understand CSR. On the one hand, identifying the impacts of the organization's decisions and activities with regard to the central issues of ISO 26000; and on the other hand, identifying stakeholders and engaging with them (Aubrun et al. 2010). The idea is to inform the company on a relevant and prioritized action plan. The latter will be built based on two fundamental input elements:

- The impacts of the company's activities by taking into consideration the perception of the life cycle on the entire value chain and the systematic consideration of seven central issues of the standard, namely: Organizational Governance, Human Rights, Labor Relations and Conditions, Environment, Fair Practices, Consumer Issues, Community and Local Development.
- Relevant stakeholder expectations.

In relation to this second practice, it is often reported in the literature that CSR is traditionally associated with stakeholder theory which serves as the basis for all its applications. The term Stakeholder refers to "*any group potentially affected by the decisions and directions of the company*" Robbins et al. (2014). This definition suggests that the stakeholder -Company relationship is one-way, yet, the Iso 9001 version 2015, added the notion of reciprocity by defining a Stakeholder as "*any person or organization that can either influence a decision or activity, or be influenced or feel influenced by a decision or activity*". The latter corroborates (Freeman 1984)'s definition of a stakeholder as "*any group or individual who can affect or be affected by the achievement of the company's objectives*".

The dissemination of this stakeholder practice via the most popular Iso standards (9001, 45001 & 14001) has contributed to its appropriation within the organizations adopting its management systems. From the perspective of this stakeholder theory, the company is socially responsible in its internal and external sphere of activity. Indeed, the company is considered as an entity and an open system having several connections with its economic, financial, regulatory, political & ecological environment. Considering the needs and expectations of stakeholders in the company's strategy provides a framework for analyzing possible connections and possible responses. Of course, not all the stakeholders are at the same level of priority. Depending on the influence of each of the stakeholders and the company's ability to respond to them, the company will evaluate the common interests that may be convergent and/or contradictory.

In concrete terms, this translates first into the creation of a map that illustrates the categorization of the different classes of stakeholders according to the nature and intensity of their interaction with the firm. In the literature, many classifications of stakeholders are available: "voluntary and involuntary" (Clarkson 1995) according to whether they have a right arising from contractual relations, in the same sense but with a different appellation (Garvare and Johansson 2010; Carroll and Buchholtz 2014) propose the categorization "primary or secondary" under this appellation are grouped the stakeholders with formal, contractual or official relations with the company. According to classifications (Garvare and Johansson 2010), stakeholders can be divided into primary and secondary stakeholder groups, as well as manifest and latent stakeholders. Primary stakeholders are those actors who provide the necessary means or support to the organization, which, if their demands are not met, they can cause consequences for the organization. These "primary stakeholders" can be customers, employees, colleagues, investors, shareholders and the government. In contrast, secondary stakeholders are those actors who do not directly provide essential means of support to the organization, but who have sufficient influence to merit consideration as more than just interested parties. These "secondary stakeholders" may include non-governmental organizations, academics, the media, fair trade organizations, environmental advocacy groups, and other individuals or organizations that in some way influence the organization.

6.2.2 Social Sustainability in SMEs

Companies worldwide, including SMEs, are recognizing that tackling social issues would result in an economically successful business performance (Medel et al. 2011). Recent research implies that more SMEs are engaging in social sustainability initiatives (Del Giudice et al. 2017), as it has become necessary to manage their social responsibility.

Despite the growing commitment of SMEs towards SS, it is still not widespread among this category of enterprises (Dillard et al. 2010). Various features of SMEs have made it challenging to reframe the socially sustainable businesses practices. First, the adoption of sustainable practices requires considerable investments in financial and human resources, however, SMEs suffer from spare resources and financial constraints (Klovienė and Speziale 2015; Siegel et al. 2019). Furthermore, even though SMEs play a major role in the economy and have the potential to easily create jobs, they can also create the opposite effect by putting an end to these jobs with each business defeat (Dillard et al. 2010). This can be explained by the fact that SMEs are easily impacted by external economic disruptions (Falkner and Hiebl 2015). Second, the existing SS standards are inappropriate for SMEs, since they have been tailored to fit the requirements of larger firms. Hence, SMEs need to have suitable standards and guidelines for SS that would match their characteristics and profiles (Klovienė and Speziale 2015). In addition, SMEs mostly implement "short term survival strategies" (Dillard et al. 2010), which do not go in line with SS practices. SMEs, generally reflect the values and ethics of their direct owner (Graafland et al. 2003), which can be a doubled edge weapon in case of the non-adherence of social values by the owner. Third, SMEs are still skeptical about the benefits that would escort the adoption of SS practices; hence, they are hesitant to spend time and resources on other issues besides business concerns (Dillard et al. 2010). Moreover, convincing customers and stakeholders to invest in social or environmental issues in the context of SMEs is not a simple deal, since SMEs lack mediatic visibility in comparison to multinational corporations (Jenkins 2006).

Despite, the challenging context of SMEs for the application of SS, it was found that the engagement of business owners in this implementation would result in the business sustainable development, given to the flexible communication and the absence of too many hierarchical levels (Rueda-Manzanares et al. 2008). SMEs can conquer a competitive advantage, through SS, only if effort is done on the development of practices that are convenient to the specific SME and its profile, not through generic strategies and assumptions.

6.2.3 Social Responsibility and SMEs: Why?

Applied to the industrial context, as detailed above, CSR relies heavily on stakeholder theory. Factors related to the relevance of and dependence on its stakeholders are decisive in the appropriation of CSR objectives in the strategy of companies. Particularly for SMEs (Quairel and Auberger 2005), explain that this type of structure is more dependent on its environment (in terms of stakeholders) than are large companies. This suggests that they are under greater pressure and therefore more inclined to adopt a CSR approach, but according to the same authors, the actors are not the same and their influence on SMEs is different and weaker than it is on large companies.

Indeed, the example of shareholder pressure in large companies is often ferocious, whereas for SMEs, the small number of shareholders, or even their absence (it may be that the manager is the only shareholder in his company), relieves them of this pressure. Unencumbered, investment in CSR actions is more affordable, but from the point of view of the owner-manager, he focuses his efforts on achieving short-term economic performance, limiting the risks of investment in other actions. As a result, the situation of the owner-manager appears, in this context, to be an obstacle to the development of a CSR policy in SMEs (Quairel and Auberger 2005).

Thus, stakeholder pressure does not have the same magnitude when it comes to small or large companies. Another example, in a country where the workforce is abundant and accessible at low cost, the pressure of this stakeholder is almost imperceptible for SMEs, especially in the absence of an organizational form such as the union. In this context, CSR will be reduced to the only moral and regulatory aspect. Similarly, when SMEs are main suppliers of large groups or multinationals, the dependence on their customers is vital. Consequently, taking their requirements into consideration will be a constraint, as it will jeopardize the SME's survival. Therefore, the pressure from stakeholders often comes from customers, especially in certain sectors such as automotive, mechanical where SMEs are subcontractors who are obliged to align themselves with the environmental and occupational health and safety requirements demanded by their group.

In any case, the reasons why business leaders will opt for responsible management can be found in the list drawn up by researchers (Robbins et al. 2014) citing the main arguments in favor of corporate social responsibility.

- Public expectation: social expectations of companies have increased considerably since the 1960s. Public opinion in favor of social objectives in addition to economic issues is clearly heard.

- Long-term benefits: socially responsible companies have more guarantees of long-term benefits. The reasons are better community relations and a better brand image.
- Ethical obligations: a company can and should have a conscience. It must prove social responsibility for its own good.
- Brand image: companies try to improve their brand image in order to attract more customers and better employees, to access financial markets, etc. Knowing that public opinion values social causes, the company can improve its image by setting objectives of this kind.
- Better environment: corporate involvement can help solve complex social problems and thereby improve the quality of life and attractiveness of the community to retain talent,
- Blocking public decision-making public regulations add economic co-payment and limit managers' decision room. By becoming socially responsible, companies can slow down the emergence of new restrictive regulations.
- Balance between responsibility and power: companies have a lot of power in society; they must develop a high level of responsibility to find a balance. When power prevails, the imbalance is a source of irresponsible behavior against the public good.
- Shareholder interests: social responsibility improves stock prices in the long term. The stock market sees socially responsible companies as less risky and less subject to public attack. It therefore rewards their actions with a higher price/earnings ratio.
- Ownership of resources: companies have the financial resources, technical experts and management talent to carry out public projects and charities.
- Prevention rather than cure: Sometimes social problems need to be addressed and companies need to address them before the solution becomes too complex and costly. To do this, efforts must be diverted momentarily from the production of goods and services.

Whatever one thinks of these various arguments, the information and digital age reminds us of order and reality. The reality that information circulates today at high "speed", catches up with the reputation of companies that sometimes find themselves in difficult situations: boycotted by their customers due to a bad social practice, sanctioned by public opinion following a work accident in which the responsibility of the company has been proven … To avoid these situations, companies have every interest in considering the social dimension in their strategic planning. Beyond the considerations already mentioned, the considering of CSR for SMEs remains in a logic of professional ethics and according to the capacities and resources available.

6.3 The Problematic of Integrating Lean, Six Sigma and Social Sustainability in SMEs

Implementing LM, LSS and SS in SMEs is faced by various barriers. SMEs lack skilled resources and funding in order to support the integration process of LM, LSS and SS (Choudhary et al. 2019). These projects require extra investments in time, capital, advanced technologies and external technical assistance (Belhadi et al. 2018). Moreover, SMEs are generally resistant to change and believe that the implementation of LM, LSS and SS is an uncertain trade, escorted with various risks (Shi et al. 2008). Furthermore, SMEs lack qualified staff that need skills, adequate training and tools to apply the practices of LM and LSS.

Indeed, Both LM and LSS can improve the pillars of sustainability including the social one (Sajan et al. 2017). Findings suggest that LM practices provide better working conditions through the improvement of workplace safety, health and organization (Minh et al. 2019). Besides, LM enhances workers skills and problem-solving ability which creates a sense of accomplishment and pride (Sajan et al. 2017; Huo et al. 2019). Finally, LM promotes employee influence, involvement and support (Järvenpää and Lanz 2020). Similarly, to LM, LSS can positively influence on the working conditions of employees. Further, according to Sony et al. (2020) LSS can enhance the health and wellbeing of society, public services and infrastructure along with quality of education. LSS is a sophisticated tool; therefore, employees who are master it, develop both confidence and a sense of achievement, besides the recognition of top management that realizes the amount of effort employees have to make in the training process (Antony 2011).

However, adding to the constraints of SMEs, LM and LSS can also have the adverse effect on SS (Sajan et al. 2017). First, customer demand and value are what drive LM and LSS; therefore, SS can be neglected in favor of customer 's orientations (Järvenpää and Lanz 2020). Besides, reaching an equilibrium between economic and social performance is not a trivial question, since there is no clear threshold of social responsibility that a firm must fulfil (Besiou and Van Wassenhove 2015). Second, LM can negatively affect workers wellbeing, as it can cause stress, fatigue and loss of autonomy due the increased workload and the repetitiveness of tasks (Martínez-Jurado and Moyano-Fuentes 2014). Finally, employees may feel untrusty and unsecure about their current job positions, given the fact that LM promotes cross training and multitasking (Nawanir et al. 2020).

In short, SMEs are called to understand the benefits, opportunities and prevailing challenges that come with LM, LSS and SS in order to successfully implement and maintain both operational and social performance. They also need to implement advanced tools that can foster the relationship between LM, LSS and SS.

6.4 Role of Smart Manufacturing to Foster the Relationship Between Lean, Six Sigma and Social Sustainability in SMEs

Nowadays, LM, LSS and SS are major occupations that need to be addressed in SMEs. The new era of smart manufacturing, enabled by the novel technologies of industry 4.0 (I4.0), has been found to positively impact the relationship between the three paradigms of LM, LSS and SS (Varela et al. 2019).

Both organizations and societies can move towards sustainable development by the means of the digital transformation (de Sousa Jabbour et al. 2018). The technologies of I4.0 will not only create numerous job opportunities in different fields such as software engineering, robotics and machine learning, but will also eliminate low skilled jobs (Brougham and Haar 2018). The new opportunities of employment will also improve the social welfare of employees, since the salaries are increased for these skilled job positions (Sung 2018). In the same vein, employee's data can be exploited in order to design career development plans and training programs suitable for each worker, which will enhance further the human resource development (Ghobakhloo 2020). This will also foster the implementation of LM and LSS programs since they require a high level of training (Kumar et al. 2006). These improvements can eventually lead to reduction of the failure rate due to the organizational changes that escort LM and LSS projects (Kolberg and Zühlke 2015).

Furthermore, technologies like the internet of things (IoT) will reduce carbon emissions thanks to its efficient and real time monitoring (Kamble et al. 2018) Robotics, artificial intelligence and machine learning will provide an improved hazard identification and accurate risk assessment (Cherubini and Navarro-Alarcon 2021). This will guarantee employee's safety and well-being. These same technologies can boost the level of LM and LSS applications in SMEs, since most companies that have an increased adoption of I4.0 applications note an improved implementation of LM practices (Tortorella and Fettermann 2018). Big data analytics (BDA) alongside with IoT will provide a better product identification, real time tracking and better visibility, which will reduce waste, improve employee decision making, and foster continuous improvement initiatives of LM (Mayr et al. 2018). Besides, LSS is "data driven" (Belhadi et al. 2020), which means it can be positively impacted by the power of BDA. This can improve the process of data collection, treatment and analysis in order to highlight the root causes of process deviations and additionally reduce the cost of LSS implementation (Arcidiacono and Pieroni 2018). Moreover, changeover times can be reduced thanks to the smart and real time communication between humans and products (Sanders et al. 2016) Finally the use of robotics can also enhance the production planning in real time and ensure the continuous flow required by LM (Wagner et al. 2017).

In a nutshell: I4.0 has a powerful ability to foster the implementation of the paradigms of LM, LSS and SS in SMEs.

6.5 Framework of Integrating Smart Manufacturing and Lean Six Sigma for Higher Social Sustainability

We propose a unique integrated framework for the integration of LM, LSS and SS supported by smart manufacturing features. As elaborated in Fig. 6.1, the base for the framework is the set of the critical success factors that would promote the implementation of the four paradigms in addition to the tools and technologies that would foster this integration.

Figure 6.1 shows that some issues can threaten the integration of LM, LSS and SS, thus the success of this integration is related to some factors. First, this integration would require highly skilled personnel, therefore SMEs are called to enhance their training programs and peruse the strength and capabilities of employees (Jaiswal and Kumar 2016). Second, this implementation needs high investments, thus SMEs should consider exploring other financing options like supply chain finance method that's "an approach for two or more organizations in a supply chain, including external service providers, to jointly create value through means of planning, steering, and controlling the flow of financial resources on an interorganizational level" (Hofmann 2005). Finally, SMEs must be ready to undertake some drastic organizational changes in order to shift their culture and achieve long term goals through the motivation and implication of its employees in this transformation (Hoffman and Henn 2008).

On the other hand, the deployment of the integrated models needs specific tools and techniques to implement and support this integration. SMEs need to consider a toolbox of social decision support tools and methods that focus on human rights, work, life quality, impact on the society and context embeddedness of social aspects (Janker and Mann 2020). One of these tools is corporate social responsibility

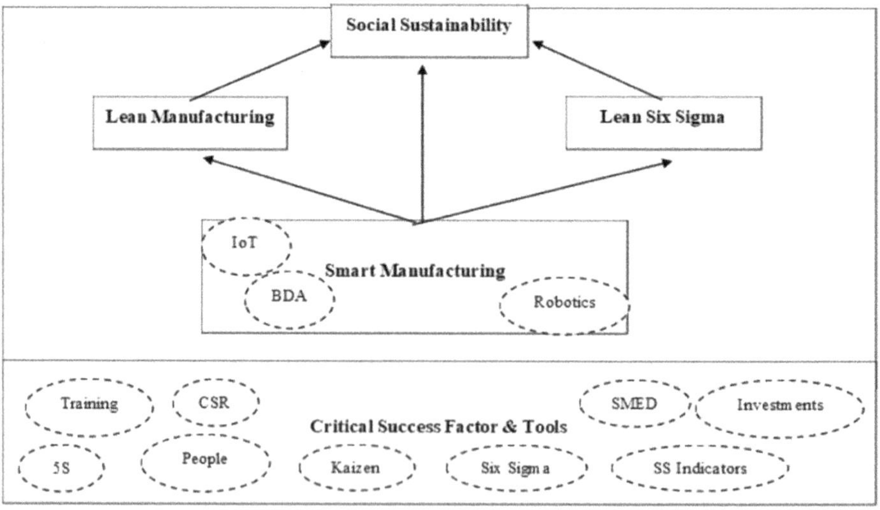

Fig. 6.1 Integrated Framework for social sustainability, lean, six sigma and smart manufacturing

(CSR) that shows the commitment of the company to enhance social and environmental issues resolution (Sutherland 2016). Furthermore, SMEs must assess their SS performance through social and socio-economic indicators like labor equity, health and safety (Hutchins and Sutherland 2008), regional hiring and Decent working conditions (Schwarz et al. 2016).

Lean six sigma puts forward a multitude of tools that can assist SMEs in reducing waste and social impacts (Wilson 2010). The tools include value stream mapping, SMED, just in time, six sigma, kaizen, statistical process control, 5S... etc., however it is necessary for SMEs to cluster these tools in a way that fits its vision and social performance (EPA 2003).

The framework considers smart manufacturing as the driven force that enables and supports the SS, LM and LSS. While also stressing on the direct impact of LM and LSS on SS. As elaborated on the previous sections, technologies like BDA, IoT and robotics have a great potential in fostering this integration. Therefore, SMEs are called to consider investing in these technologies and selecting the one and/or more in order to succeed in its quest for SS.

The four paradigms will work together as a powerful tool to support SMEs in assessing their strengths, weaknesses and prospects for improvement. Furthermore, SMEs will benefit from each paradigm individually and collectively to achieve sustainable performance. Finally, the set of tools and critical success factors will enable SMEs to build a strong, effective and efficient system that further empowers continuous improvement and SS.

6.6 Conclusion

The new era of SM, enabled by the novel technologies of I4.0 has been found to positively impact the relationship between the three paradigms of LM, LSS and SS (Varela et al. 2019).

In this chapter, we further highlight the role of SM for fostering and overcoming the challenges that face this implementation in the context of SMEs. Through the development of holistic framework, we hope that SMEs could consider the adoption of I4.0 technologies in order to improve both operational and social performance.

References

Ajmal MM, Khan M, Hussain M, Helo P (2018) Conceptualizing and incorporating social sustainability in the business world. Int J Sustain Dev World Ecol 25(4). Taylor & Francis: 327–339

Antony J (2011) Six sigma vs lean. Int J Prod Perform Manag. Emerald Group Publishing Limited

Arcidiacono G, Pieroni A (2018) The revolution lean six sigma 4.0. Int J Adv Sci Eng Inf Technol 8(1):141–149. https://doi.org/10.18517/ijaseit.8.1.4593

Aubrun M, Bermond F, Brun E, Cortot J-L, Delchet-Cochet K, Graffin O et al (2010) ISO 26000 responsabilité sociétale: comprendre, déployer, évaluer. AFNOR, Paris

Belhadi A, Kamble SS, Zkik K, Cherrafi A, Touriki FE (2020) The integrated effect of big data analytics, lean six sigma and green manufacturing on the environmental performance of manufacturing companies: the case of North Africa. J Clean Prod 252. https://doi.org/10.1016/j.jclepro.2019.119903

Belhadi A, Sha'ri YBM, Touriki FE, El Fezazi S (2018) Lean production in SMEs: literature review and reflection on future challenges. J Ind Prod Eng 35(6):368–382

Besiou M and Van Wassenhove LN (2015) Addressing the challenge of modeling for decision-making in socially responsible operations. Prod Oper Manag 24(9), Wiley Online Library: 1390–1401

Brougham D, Haar J (2018) Smart technology, artificial intelligence, robotics, and algorithms (STARA): employees' perceptions of our future workplace. J Manag Organ 24(2). Cambridge University Press: 239–257

Cadet I (2014) Responsabilité sociale de l'entreprise (RSE), responsabilité éthiques et utopies, Les fondements normatifs de la RSE, Etude de la place du droit dans les organisations. Conservatoire national des arts et métiers

Capron M, Quairel-Lanoizelée F (2004) Mythes et réalités de l'entreprise responsable: acteurs-enjeux-stratégies. Alternatives économiques, La Découverte, Paris

Carroll AB, Buchholtz AK (2014) Business and society: ethics, sustainability and stakeholder management. CENGAGE Learning

Carroll AB (1991) The pyramid of corporate social responsibility: toward the moral management of organizational stakeholders. Bus HorizS 34(4). Amsterdam: 39–48

Cherubini A, Navarro-Alarcon D (2021) Sensor-based control for collaborative robots: fundamentals, challenges, and opportunities. Front Neurorobotics 14. Frontiers: 113

Choudhary S, Nayak R, Dora M, Mishra N, Ghadge A (2019) An integrated lean and green approach for improving sustainability performance: a case study of a packaging manufacturing SME in the UK. Prod Plan Control 30(5–6). Taylor & Francis: 353–368

Clarkson ME (1995) A stakeholder framework for analyzing and evaluating corporate social performance. Acad Manag Rev 20(1):92–117

de Sousa Jabbour ABL, Jabbour CJC, Foropon C, Godinho Filho M (2018) When titans meet—can industry 4.0 revolutionise the environmentally-sustainable manufacturing wave? The role of critical success factors. Technol Forecast Soc Chang 132 (January). Elsevier: 18–25. doi:https://doi.org/10.1016/j.techfore.2018.01.017

Dillard J, Pullman ME, Loucks ES, Martens ML, Cho CH (2010) Engaging small-and medium-sized businesses in sustainability. Sustain Account Manag Policy J. Emerald Group Publishing Limited

EPA (2003) Lean manufacturing and the environment: research on advanced manufacturing systems and the environment and recommendations for leveraging better environmental performance. http://www.epa.gov/lean/environment/pdf/leanreport.pdf. Accessed 10 July 2015

Falkner EM, Hiebl MRW (2015) Risk management in SMEs: a systematic review of available evidence. J Risk Financ. Emerald Group Publishing Limited

Forman B, Gey J-M, Bonnifet F (2007) Qualité sécurité environnement: construire un système de management intégré. AFNOR

Freeman RE (1984) Strategic management: a stakeholder approach. Pitman, Boston

Garvare R, Johansson P (2010) Management for sustainability—a stakeholder theory. Total Qual Manag Bus Excel 21(7): 737–744

Ghobakhloo M (2020) Industry 4.0, digitization, and opportunities for sustainability. J Clean Prod 252. Elsevier: 119869

Del Giudice M, Khan Z, De Silva M, Scuotto V, Caputo F, Carayannis E (2017) The microlevel actions undertaken by owner-managers in improving the sustainability practices of cultural and creative small and medium enterprises: a United Kingdom–Italy comparison. J Organ Behav 38(9). Wiley Online Library: 1396–1414

Graafland J, Van de Ven B, Stoffele N (2003) Strategies and instruments for organising CSR by small and large businesses in the Netherlands. J Bus Ethics 47(1). Springer: 45–60
Hoffman AJ, Henn R (2008) Overcoming the social and psychological barriers to green building. Organ Environ 21(4). Sage Publications Sage CA: Los Angeles, CA: 390–419
Hofmann E (2005) Supply chain finance: some conceptual insights. Beiträge Zu Beschaffung Und Logistik. Springer: 203–214
Hollander R, Amekudzi-Kennedy A, Bell S, Benya F, Davidson C, Farkos C, Fasenfest D et al (2016) Network priorities for social sustainability research and education: memorandum of the integrated network on social sustainability research group. Sustain Sci Pract Policy 12 (1). doi:https://doi.org/10.1080/15487733.2016.11908150
Huo M-L, Boxall P, Cheung GW (2019) Lean production, work intensification and employee well-being: can line-manager support make a difference? Econ Ind Democr. SAGE Publications Sage UK: London, England, 0143831X19890678
Hutchins MJ, Sutherland JW (2008) An exploration of measures of social sustainability and their application to supply chain decisions. J Clean Prod 16(15). Elsevier: 1688–1698
Jaiswal P, Kumar A (2016) Analyzing barriers of lean manufacturing adoption in Indian SMEs using an integrated approach of grey decision making trial and evaluation laboratory (DEMATEL). Proc World Congr Eng 2:688–691
Janker J, Mann S (2020) Understanding the social dimension of sustainability in agriculture: a critical review of sustainability assessment tools. Environ Dev Sustain 22(3). Springer: 1671–1691
Järvenpää E, Lanz M (2020) Lean manufacturing and sustainable development. Responsible Consum Prod. Springer: 423–432
Jenkins H (2006) Small business champions for corporate social responsibility. J Bus Ethics 67(3). Springer: 241–256
Kamble SS, Gunasekaran A, Gawankar SA (2018) Sustainable industry 4.0 framework: a systematic literature review identifying the current trends and future perspectives. Process Saf Environ Prot 117. Elsevier: 408–425
Klovienė L, Speziale M-T (2015) Is performance measurement system going towards sustainability in SMEs? Proc Soc Behav Sci 213:328–333. https://doi.org/10.1016/j.sbspro.2015.11.546
Kolberg D, Zühlke D (2015) Lean automation enabled by industry 4.0 technologies. IFAC-PapersOnLine 28(3):1870–1875. https://doi.org/10.1016/j.ifacol.2015.06.359
Koning J (2002) "More on MOST:" National Unesco Commission, Amesterdam, no. September: 63–90
Kumar M, Antony J, Singh RK, Tiwari MK, Perry D (2006) Implementing the lean sigma framework in an Indian SME: a case study. Prod Plan Control 17(4). Taylor & Francis: 407–423
Littig B, Griessler E (2005) Social sustainability: a catchword between political pragmatism and social theory. Int J Sustain Dev 8(1–2). Inderscience Publishers: 65–79
Martínez-Jurado PJ and Moyano-Fuentes J (2014) Lean management, supply chain management and sustainability: a literature review. J Clean Prod 85. Elsevier Ltd: 134–150. doi:https://doi.org/10.1016/j.jclepro.2013.09.042
Mayr A, Weigelt M, Kühl A, Grimm S, Erll A, Potzel M, Franke J (2018) Lean 4.0—a conceptual conjunction of lean management and industry 4.0. Proc CIRP 72. Elsevier B.V.: 622–628. doi:https://doi.org/10.1016/j.procir.2018.03.292
Medel F, García L, Enriquez S, Anido M (2011) Reporting models for corporate sustainability in SMEs. Inf Technol Environ Eng, 407–418. Springer
Minh KS, Zailani S, Iranmanesh M, Heidari S (2019) Do lean manufacturing practices have negative impact on job satisfaction? Int J Lean Six Sigma. Emerald Publishing Limited
Nawanir G, Lim KT, Lee KL, Moshood TD, Ahmad ANA (2020) Less for more: The structural effects of lean manufacturing practices on sustainability of manufacturing SMEs in Malaysia. Int J Supply Chain Manag 2(2):961–975
Quairel F, Auberger M-N (2005) Management responsable et PME: une relecture du concept de "Responsabilité sociétale de l'entreprise". La revue des Sciences de Gestion 40(211/212):111

Robbins S, Decenzo D, Coulter M, Rüling C-C (2014) Management: l'essentiel des concpets et pratiques, 9th édn. Nouveaux Horizons, France
Rueda-Manzanares, A, Aragón-Correa JA, Sanjay Sharma (2008) The influence of stakeholders on the environmental strategy of service firms: the moderating effects of complexity, uncertainty and munificence. Br J Manag 19(2). Wiley Online Library: 185–203
Sajan MP, Shalij PR, Ramesh A (2017) Lean manufacturing practices in Indian manufacturing SMEs and their effect on sustainability performance. J Manuf Technol Manag. Emerald Publishing Limited
Sanders A, Elangeswaran C, Wulfsberg J (2016) Industry 4.0 implies lean manufacturing: research activities in Industry 4.0 function as enablers for lean manufacturing. J Ind Eng Manag 9(3):811–833. https://doi.org/10.3926/jiem.1940
Schwarz J, Schuster M, Annaert B, Maertens M, Mathijs E (2016) Sustainability of global and local food value chains: an empirical comparison of Peruvian and Belgian Asparagus. Sustainability 8(4). Multidisciplinary Digital Publishing Institute: 344a
Shi H, Peng SZ, Liu Y, Zhong P (2008) Barriers to the implementation of cleaner production in Chinese SMEs: government, industry and expert stakeholders' perspectives. J Clean Prod 16(7). Elsevier: 842–852
Siegel R, Antony J, Garza-Reyes JA, Cherrafi A, Lameijer B (2019) Integrated green lean approach and sustainability for SMEs: from literature review to a conceptual framework. J Clean Prod 240. Elsevier: 118205
Sloan TW (2010) Measuring the sustainability of global supply chains: current practices and future directions. J Glob Bus Manag 6(1):92–107. http://www.faculty.uml.edu/tsloan/papers/Sloan-JGBM-v6n1-2010.pdf
Sony M, Naik S, Antony J (2020) Lean six sigma and social performance: a review and synthesis of current evidence. Q Manag J 27(1). Taylor & Francis: 21–36
Sung, TK (2018) Industry 4.0: a Korea perspective. Technol Forecast Soc Chang 132. Elsevier: 40–45
Suppliers DC (n.d.) Social sustainability in developing country suppliers: an exploratory study in the ready made garments industry of Bangladesh
Sutherland E (2016) Corporate social responsibility: the case of the telecommunications sector. Info. Emerald Group Publishing Limited
Tortorella GL, Fettermann D (2018) Implementation of industry 4.0 and lean production in Brazilian manufacturing companies. Int J Prod Res 56(8). Taylor & Francis: 2975–2987. doi:https://doi.org/10.1080/00207543.2017.1391420
Vallance, S, Perkins HC, Dixon JE (2011) What is social sustainability? A clarification of concepts. Geoforum 42(3). Elsevier Ltd: 342–348. doi:https://doi.org/10.1016/j.geoforum.2011.01.002
Vallejo, VF, Antony J, Douglas JA, Alexander P, Sony M (2020) Development of a roadmap for lean six sigma implementation and sustainability in a Scottish packing company. The TQM Journal. Emerald Publishing Limited
Varela L, Araújo A, Ávila P, Castro H, Putnik G (2019) Evaluation of the relation between lean manufacturing, industry 4.0, and sustainability. Sustainability 11(5). Multidisciplinary Digital Publishing Institute: 1439
Wagner T, Herrmann C, Thiede S (2017) Industry 4.0 impacts on lean production systems. Proc CIRP 63:125–131. https://doi.org/10.1016/j.procir.2017.02.041
Weingaertner C, Moberg Å (2014) Exploring social sustainability: learning from perspectives on urban development and companies and products. Sustain Dev 22(2). Wiley Online Library: 122–133
Wilson L (2010) How to implement lean manufacturing. McGraw-Hill Education
Yedder M-B, Farhoud M (2009) Le développement durable est-il bienvenu dans les organisations? Cas de l'implantation d'un Système de Management Environnemental en Tunisie. Développement durable & territoire

Chapter 7
How to Measure Progress in Sustainable Excellence of SMEs?

Objectives

1. Explain the concept of operational, environmental and social performance,
2. Present the current assessment methods of sustainable excellence.

7.1 Introduction

Initially considered from a financial perspective, performance evaluation has been progressively broadened in recent years to include the notion of "societal responsibility" in strong relation to the concerns of stakeholders. This expanded "Sustainable" performance combines several dimensions at once: economic, social and environmental (Triple Bottom Line of Elkington [1997]). This is what also confirms (Quairel 2005), specifying that performance today could not be treated only from the financial angle, it is still necessary to include performance related to sustainable development strategies.

In this perspective, the performance of companies is no longer unidimensional but is a multidimensional construct (Tahon 2003; Maurel and Tensaout 2014) or "global". This is defined according to Baret (2006) as "the aggregation of economic, social and environmental performance".

However, performance evaluation is not an end but a means of enabling the company to be part of a continuous improvement process based on measurement. Therefore, measuring performance in a systematic way is an essential tool for steering companies, projects and teams. Therefore, for good governance, several questions can be raised: what are the performance indicators to adopt in the framework of Sustainable Excellence? How can they be identified and designed? What measurement systems and dashboards are available to support this activity? Are there any reference standards for the evaluation of economic, social and environmental performance? To do this, this chapter will look at the composition of "Sustainable" performance, its evaluation, its indicators and its representation.

F. E. Touriki et al., *Sustainable Excellence in Small and Medium Sized Enterprises*, Industrial Ecology, https://doi.org/10.1007/978-981-19-0371-7_7

7.2 The Notion of "Performance" in the Context of Sustainable Excellence

7.2.1 *Performance: A Multidimensional "Construct"*

According to the standard (ISO 9000 2015), performance "is a measurable result, which can relate to quantitative or qualitative findings, and relates to activities, processes, systems as well as products and services", this is a first international and standardized definition that allows to bypass the concept of performance. However, according to some authors (Saulpic et al. 2011), the definition of performance is not necessarily universal, it is a "construct" that is influenced by different factors: the type of organization, the sector of activity, the strategy pursued and the configuration of stakeholders. Thus, the multiplicity of these factors generally makes the definition of performance specific to each organization. This perspective holds some truth, insofar as the objectives and strategies differ from one company to another, so performance may have different connotations in each organization. As a result, it is constructed according to the objectives to be achieved, the organizational culture and the expectations of stakeholders.

Other authors place performance within a resilience framework such as Hollnagel et al. (2006), who explain that performance is the intrinsic capacity of the company to adapt to changes and return to a stable state. Such an ability is of crucial importance especially in crisis situations like the COVID 2019 pandemic.

Furthermore, the notion of performance is necessarily linked to a monitoring and evaluation process. This is a process through which performance indicators are measured, making it possible to assess the functioning and results of an activity in order to subsequently determine what corrective and improvement actions are possible. Indeed, Robbins et al. (2014) explain that the evaluation process can be summarized as follows: (1) The measurement of performance, (2) The comparison of achieved performance against benchmark standards, (3) The managerial action to correct deviations. These three steps are essential, especially the measurement step, which is the trigger for any improvement, because without measurement, progress will be arbitrary. We can add another preliminary step, that of "identifying indicators" which, in our opinion, is the most decisive step in this evaluation process.

7.2.2 *The Dashboards*

According to the above, evaluation is inevitably done through measurement systems, or dashboards that refer to an information medium grouping together a set of indicators over a given period and that help the management of the company and its processes. Their advantage lies in the fact that they translate performance into concrete terms while facilitating understanding and communication (Saulpic et al. 2011). However, their construction remains a rather delicate exercise to maneuver:

first, the presence of a multitude of choices of measurement axes amplifies the number and nature of the corresponding performance indicators. Indeed, these indicators must reconcile the short and the long term, the qualitative and the quantitative, the local and the global, the collective and the individual interest… These different indicators must also provide sufficient information on the company's activity or its processes while focusing on its strategic objectives. This exercise requires both transversal management (harmonization between processes in terms of measurement axes) and vertical management (harmonization between processes and strategic axes). Secondly, care must be taken in the design, development and implementation of the performance measurement system, which remains an ongoing process and not a one-off activity (Kamble et al. 2020).

In the literature, the Balanced Scorecard (BSC) remains the most widely used management tool for measuring performance. American authors (Kaplan and Norton 1992) developed a model characterized by the use of a multidimensional vision of performance that was no longer necessarily financial. In its original conception, the BSC is organized according to four perspectives that have the company's strategy and objectives as their common point of departure: (**1**) "**Financial Results**" perspective (measuring the level and evolution of financial performance); (**2**) "**Customer/Market**" perspective (includes indicators that allow the evaluation of the degree of customer satisfaction in terms of cost, deadlines and quality); (**3**) "**Internal Processes**" perspective (includes indicators of the mastery of operational processes as being determining factors for the acquisition of competitive advantages. Companies are called upon to identify their technical skills and capabilities in terms of mastered technologies and which are the key processes that must be taken into account as a priority to excel in their field); (**4**) "**Resources and Organizational Learning**" perspective (Companies must question the ways of managing human resources, knowledge and skills to achieve strategic objectives).

Since then, the BSC has been considered a strategic management system and a means of clarifying and communicating strategy (Kaplan and Norton 1996). Its construction leads to the selection and structure of information relating to the global and operational performance of the company, which is achieved by retaining a limited and complete number of indicators for each perspective. Obviously, these indicators can be reviewed and renewed according to the evolution of the internal and external environment, in order to maintain the quality of the decision and intervention. The correlation of the indicators with the strategy made it possible to link the indicators in a series of "Cause and Effect" relationships also called "Strategy Map" (Bouamama 2015).

According to Bieker and Gminder (2002), the BSC or Balanced Scorecard attempts to strike a balance in several respects:

- Goals related to the short and long term should be considered equally,
- Internal aspects (such as processes and organizational development) as well as external aspects (customers and shareholders) are taken into consideration,
- The "Financial Results" perspective, which includes indicators (lagging or ex-post), is used to describe the effect of objectives and actions already taken, while

the other perspectives include other (leading) indicators that are considered as "enablers" to get closer to the desired performance in the future.

- A balance between quantitative, i.e. financial, and qualitative non-financial facts is proposed to be maintained.

7.2.3 Sustainability Balanced Scorecard

From the outset, the idea of BSC seems consistent with sustainable performance in the sense that this dashboard already takes into consideration indicators related to "Stakeholders", such as customers, shareholders and employees. Even if this consideration is only partial due to the explicit absence of the terms "Environment" and "Society", the spirit is still present. It is conveyed through a dynamic that mobilizes the entire company in which the satisfaction of employees' expectations through their involvement and development of their skills, will lead to excellence in operational processes (quality, productivity, availability…), which would promote customer satisfaction and consequently that of shareholders.

However, the growing interest in environmental and social issues has led the scientific and professional community to move towards more advanced measurement models than the Sustainability Balanced Scorecard (SBSC) (Bieker and Gminder 2002; Zingales and Hockerts 2003). In fact, the SBSC differs from the BSC in its architecture by explicitly recognizing the objectives and performance measures related to sustainable excellence (Hansen and Schaltegger 2014). This architecture refers directly to the content question: should we add other additional perspectives entitled "Environment", "Society", or introduce indicators of these same dimensions into the 4 classical perspectives of Kaplan and Norton. Indeed, these questions have raised debates in the scientific community (Bieker and Gminder 2002; Figge et al. 2002), these authors propose at least three possibilities or modalities for integrating the dimensions of sustainable development into the BSC:

1. **Partial integration**: one or two "social and environmental" objectives and indicators are placed in the perspectives that are most relevant or have the greatest impact on the sustainability of the company.
2. **Additive integration**: consolidates the environmental and social aspects into a separate 5th perspective.
3. **Total integration**: environmental and social aspects are integrated into all perspectives of the BSC. This is the most extensive and complete approach allowing for a real and effective consideration of the different dimensions of sustainable development.

For example, the founders of BSC suggest extending the "Customer" perspective to all the partners of the company (Kaplan and Norton 2001). Environmental and social indicators can be attached to "Internal Processes" and "Resources and Organizational Learning" respectively. In any case, Bieker and Gminder (2002) point out that the design and choice of indicators depend on perfect collaboration between the management team and other departments. The weight of each perspective and its corresponding indicators is intimately linked to the weight allocated to this perspective in the strategy and objectives of each company.

The SBSC offers the advantage of specifying the measurement axes, however, the choice of indicators for each axis remains a challenge for any organization. In this sense, the research carried out in the framework of sustainable excellence gives rise to measurement systems specifying performance indicators that can serve as a basis for the construction of a "balanced scorecard" (Fercoq 2014). The latter synthesizes the indicators underlying several models for assessing sustainable performance (Table 7.1).

Sustainable performance is thus apprehended from economic performance (via financial and operational indicators), environmental performance (including resource consumption and waste production), while social performance is assessed according to indicators related to motivation and management, and societal performance in relation to the main stakeholders (customers and suppliers).

It is also noted that the performance indicators in the framework of sustainable excellence concern practically all the activities of the company and refer to the objectives of the organization, the regulations and the stakeholders. It is true that in most cases the indicators can be deduced from the specific objectives and expressed in a tangible and measurable way. However, in some cases, the performance indicator cannot be defined in quantifiable terms. Managers have to resort to subjective measures. These have important limitations, but they are still better than a complete lack of reference and control (Robbins et al. 2014).

It should be noted that, in general, performance indicators refer to specific and recorded information on the functioning of the company's various processes. According to AFNOR (2000), "selected information, associated with a criterion, intended to observe changes at defined intervals". According to this definition, the indicators are known by their temporal character which must be observed and measured regularly. They must therefore provide the necessary information in a timely manner. There is a need for continuous review of performance measures, as they are constantly changing in response to the changing internal and external environment.

Also, these performance indicators are meaningful information with a particular meaning for their recipients. As Lorino (2003) points out, an indicator is information that should help an individual or collective actor to steer the course of an action towards the achievement of an objective, or that should enable him or her to evaluate the result. Their role is therefore both reactive, by setting up actions after the fact to deal with deviations and malfunctions that have been observed, and preventive, by taking certain decisions in order to anticipate dreadful events (particularly for alert-type indicators, which draw the attention of the recipient to the need to take the necessary actions quickly in order to avoid an undesirable situation (e.g., indicator on the minimum stock).

Finally, in the context of Industry 4.0, it is fundamental to converge towards an intelligent measurement system based on the analysis of Big Data and access to

Table 7.1 Performance indicators for a balanced scorecard

Economic performance	Environmental performance (waste related)	Societal performance (employee-centric)	Societal performance (customers, suppliers, community)
Financial indicators	**Ratios**	**Indicators related to motivation**	**Customer-related indicators**
• Net profit—earnings (b, c, d, f) • Return on capital employed (b, c, f) • Earnings per share (c, f) • Operating values ©	• Resource (material) productivity (b, f) • Non-quality rate (f)	• Staff satisfaction survey (c) • Absenteeism / Lateness (a, c, d) • Health and safety (expenditure on prevention, accident rate) (a, c, d, e, f, i) • Productivity and quality (a, d, f, h) • Number of suggestions per person annually (a) • Turnover per year (a, d, f, i)	• Customer satisfaction index: service rate, product quality, communication, labeling (b, c, f, i) • Customer retention rate (c) • Consumer health and safety (c, i)
Operational indicators	**Inputs/outputs**	**Indicators related to management**	**Indicators related to suppliers**
• Turnover (c, g) • Market share growth rate (c, h) • Rate of sales generated by new products (c) • Costs (wages, materials, energy) (a, c, d, e, f, h) • Productivity (a, c, d, h) • Quality (a, c, h) • Investment/area utilization rate (c, g) • Lead time (production) / Cycle time (a, c, g) • Time To Market (c) • Inventory turnover rate (c)	• Material consumption by weight or volume (i) • Consumption of toxic materials (g, h) • Percentage of materials consumed that are of recycled origin (i) • Recycling rate (components, materials, energy)—incinerated waste rate (d, e, f, g, h) • Waste management costs: rate of waste and scrap per product, volume of waste treated, internal management costs (b)	• Number of hierarchical levels (d) • Training expenses (f) • Talent retention rate (c) • Length of service (f) • Percentage of women in senior management (d, f) • Ratio of male to female base salary by employee category (i) • Diversity (d, f)	• Supplier satisfaction index: service rate, quality of supply (f) • Supplier loyalty rate (f)

(continued)

Table 7.1 (continued)

Economic performance	Environmental performance (waste related)	Societal performance (employee-centric)	Societal performance (customers, suppliers, community)
	Prevention		**Indicators related to the community**
	• Prevention expenditures: training, compliance with rules and standards (f, g) • Rate of ISO 14001 certified suppliers (e, g, h)		• Amount of donations, sponsorship operations (d, f)
(a) Suzaki, 1993; (b) Kainuma and Tawara, 2006; (c) Bhasin, 2008; (d) O'Connor et Spangenberg, 2008; (e) Cuthbertson R. et Piotrowicz W., 2011; (f) Boon Heng et al., 2012; (g) Hajmohammad et al.,2013; (h) Hasan M., 2013; (i) Global Reporting Initiative, 2013			

Source Fercoq (2014)

information in real time. This in order to improve the reactivity of the company and develop its proactive aspect (Kamble et al. 2020).

7.3 Operational Performance

Operational performance is linked to indicators of the results or functioning of the company's operational or business processes. According to Morin et al. (1996), the results of indicators related to productivity, profitability and quality are critical for the stability and growth of the company in its environment, in the same way, these indicators bring satisfaction to the interests of shareholders and the concerns of managers. Very often, these performances are evaluated through objectives and indicators articulated according to three criteria: cost (to be controlled), time (to be optimized) and quality (to be perpetuated) (Tahon 2003). According to the same author, a fourth criterion can be added to the list, that of flexibility, i.e. the company's ability to deal with the hazards/fluctuations of customer demand. In other words, operational performance seems to be above all a question of optimization, with maximization of results and minimization of costs as factors.

In many cases, performance indicators in the company try to cover two aspects of the management of the different processes: one aspect related to results and another related to operations. In this sense, two categories of indicators can be distinguished:

- **Result indicators:** are directly linked to the efficiency of the process. They measure the degree to which the objectives of an activity or process have been achieved through what has been planned and what has been done. It can also be a question of efficiency when the means and resources for carrying out the activity are taken into account in the calculation of the indicator.

- **Operating indicators**: serve as benchmarks for action and timely decisions to avoid the worsening of a given situation.

Some indicators can be both operating and result indicators. For example: the synthetic efficiency rate is an operating indicator when it is used to provide information on the real availability of an equipment or a machine (availability rate). It is also a result indicator in the sense that it is used to control production through the performance and quality rate (influenced by defects).

It should be noted that the measurement phase that triggers any performance assessment process requires methodologies and tools to help decision-makers either to confirm the results of a project, or to take stock of the situation at the start of a project. As an example, in the context of a lean transformation project, many authors have provided a large number of models, indicators and tools for performance assessment and change (Belhadi et al. 2018). However, SMEs continue to behave in a very reactive manner resulting in relying on a short-term vision (Garengo and Bernardi 2007). This is due, according to several authors, to the absence of a simple and adapted method of evaluating changes towards lean production that would be perfectly adapted to SMEs. In fact, all the lean production evaluation frameworks provided by the literature have been designed and confirmed for large companies (Cocca and Alberti 2010).

For this reason, in what follows, a model for evaluating operational performance will be developed, adapted to the characteristics of SMEs. It is a simple and comprehensive model voluntarily using a number of indicators reduced to a minimum, focusing on the simplest and most quantitative indicators (Belhadi et al. 2018).

7.3.1 Indicator Identification

It is agreed that all performance indicators are derived from an objective to be achieved or a reference (Tahon 2003). Therefore, in order to identify operational indicators, it is first necessary to know the general objectives pursued by a lean approach. In this sense and according to the literature (Karlsson and Åhlström 1996; Åhlström 1998), the most common principles that represent the objectives of lean manufacturing are (1) Elimination of waste, (2) Continuous improvement, (3) multi-functional teams, (4) Zero defects, (5) Just-in-time delivery, (6) Information systems.

At the same time, it has been found that these principles are used as cornerstones for the design and development of frameworks and evaluation models for the determination of the lean level of the company "Leanness level". Also, and based on the work of some authors (Karlsson and Åhlström 1996; Pakdil and Leonard 2014; Sánchez and Pérez 2001; Duque and Cadavid 2007), it has been grouped a set of performance indicators that are most commonly used when implementing lean (Table 7.2).

Table 7.2 The primary list of lean production indicators empirically examined in lean literature

Lean principles	Indicator	Change	1	2	3	4
Elimination of waste	• Set-up time per unit		x	x		x
	• Changeover time/total production time			x	x	
	• Total down time/total machine time		x	x		x
	• Takt time/cycle time			x		
	• Manufacturing lead time			x		
	• Preventive maintenance time/total maintenance time				x	
	• Number of parts (trips) transported * Distance					x
Continuous improvement	• Number of suggestions per employee and year		x	x	x	
	• Total number of implemented suggestions/total suggestions		x	x	x	x
	• Total number of problem solving teams/total employees			x		
	• Total number of improvement activities organized		x			
Multifunctional teams	• Percentage of employees working in teams		x		x	
	• Average frequency of task rotation		x		x	
	• Total number of employees capable of assignment rotation/total employees					x
	• Total number of employees working in teams/total employees		x		x	
	• Total number of job classifications/total employees		x	x		
Zero defects	• Defect rate		x	x	x	x
	• Total defectives costs/total sales			x		
	• Total number of people dedicated primarily to quality control/total employees		x	x	x	
	• Percentage of inspection carried out by autonomous defect control		x	x	x	x
Just-in-time delivery	• Total work in progress/total sales		x	x	x	x
	• Total number of orders delivered just in time per year/total number of deliveries per year		x		x	
	• Average production and delivery lot sizes			x	x	x

(continued)

Table 7.2 (continued)

Lean principles	Indicator	Change	1	2	3	4
	• % of the total annual value or throughput of the system that is scheduled through pull mechanisms		x			x
	• Average total number of days from orders received to delivery			x		x
Information systems	• The frequency with which information is given to employees		x			x
	• Percentage of procedures that are documented in the company					x
	• Number of different measures used to assess the performance of the teams		x			

Note The indicator should increase, the indicator should decrease
(1) Karlsson and Åhlström (1996), (2) Pakdil and Leonard (2014), (3) Sánchez and Pérez (2001), (4) Duque and Cadavid (2007)

As the table shows, there are many lean indicators to measure the progress of each lean principle. Some are based on time measurement (such as setup time, change time, lead time, cycle time, down time…), while others are expressed as a percentage (defect rate, percentage of employees working in teams). In summary, a total of 28 indicators are identified.

7.3.2 Example of an Operational Performance Assessment Tool

Based on the survey conducted among SMEs (Belhadi et al. 2018), the number of indicators was reduced from 28 to 12 key indicators and that the survey revealed that they are the most used in industrial practices (Table 7.3).

To illustrate how these indicators are calculated and evaluated, an example is given in Table 7.4.

The Leanness Score is calculated for each principle as follows:

$$h_i = 1 - \frac{\sum_{j=1}^{j=n} (-1)^c \frac{(x_{ij} - x_{ij0})}{x_{ij0}}}{n}$$

Équation 1 Leanness Score
Where:

- h_i: leaness score of principle i
- n: number of indicators to evaluate the lean principle i
- x_{ij}: the value of the lean indicator

Table 7.3 The final list of the most useful lean indicators in SMEs investigated

Lean principle	Ref	Indicator
Elimination of waste	X11	Total down time/total machine time
	X12	Takt time/cycle time
	X13	Manufacturing lead time
Continuous improvement	X21	Total number of improvement activities organized
	X22	Total number of problem solving teams/total employees
Multifunctional teams	X31	Percentage of employees working in teams
Zero defects	X41	Defect rate
	X42	Total defectives costs/total sales
Just-in-time delivery	X51	Total number of orders delivered just in time per year/total number of deliveries per year
	X52	Total work in progress/total sales
Information systems	X61	Percentage of procedures that are documented in the company
	X62	The frequency with which information is given to employees

Table 7.4 Example of leanness score of lean principles calculation

	Lean principle	Indicator	Indicator value	Indicator target	Coefficient *c*		hi
Elimination of waste	X11	1.9%	1%	0	0.10		
	X12	0.74	1	1	0.74		
	X13	7.92 days	5 days	0	0.42		
			Score h1			**0.42**	
Continuous improvement	X21	31%	90%	1	0.34		
	X22	45%	60%	1	0.75		
			Score h2			**0.55**	
Multifunctional teams	X31	62%	80%	1	0.78		
			Score h3			**0.78**	
Zero defects	X41	3.07%	2%	0	0.47		
	X42	2.12%	2%	0	0.59		
			Score h4			**0.53**	
Just-in-time delivery	X51	52%	95%	1	0.55		
	X52	0.07	0.09	1	0.78		
			Score h5			**0.66**	
Information systems	X61	30%	60%	1	0.5		
	X62	4 days	2.5 days	0	0.4		
				Score h6			**0.45**

- x_{ij0}: the target of the indicator x_{ij}
- c: coefficient equal to 1 if the indicator should increase or 0 if the indicator should decrease.

7.4 Environmental Performance

Environmental performance refers to the results obtained from information related to the input and output elements of a production process. In this sense, the ISO 14031 standard on environmental performance evaluation (EPE) used the material flow model as the basis for indicators at the operational level (Jasch 2000).

Also, environmental performance is defined according to the standard (ISO 14001 2015), as a measurable result related to the management of environmental aspects. The note specifies that the results can be measured in relation to (1) the environmental policy of the organization, (2) the environmental objectives, (3) other criteria, by means of indicators.

The choice of indicators will always be an essential exercise for any company. To help in the identification and choice, several world organizations propose environmental performance indicators. For example, the Organization for Economic Cooperation and Development (OECD 2011) identifies 18 indicators to measure the impacts of industrial activities on the environment in industrialized countries. The categories of performance indicators proposed, is directly related to environmental aspects. These are defined by the (ISO 14001 2015) as elements of activities, products or services of an organization, interacting or likely to interact with the environment. Starting from the idea that any process involves consumption and manipulation of materials and energy, which leads to the production of products, services and waste. These physical transactions constitute the most direct relationship between companies and the environment (Tyteca et al. 2002) (Table 7.5).

7.4.1 Identification of Indicators from the GRI

In the context of identifying indicators to evaluate environmental performance, several authors refer to the use of those proposed by GRI (Global Reporting Indicators) (ADEME 2008; Nikolaeva and Bicho 2011; Manetti and Becatti 2009; Brown et al. 2009). These are guidelines that constitute a common framework at the international level for the elaboration of sustainable development reports (ADEME 2008). It proposes a set of indicators that allow companies to select those most appropriate to their strategy.

Originally intended for environmental reporting, the GRI has evolved to include the other dimensions of the "Triple Bottom Line (Elkington 1997)": social performance (indicators such as labor conditions and human rights), economic performance (economic impact on customers, suppliers, employees, capital providers, and

Table 7.5 Overview of the OECD sustainable manufacturing indicators

Inputs	Operations	Products
I1 Non-renewable materials intensity	O1 Water intensity	P1 Recycled/reused content
I2 Restricted substances intensity	O2 Energy intensity	P2 Recycability
I3 Recycled/ Reused content	O3 Renewable proportion of energy	P3 Renewable materials content
	O4 Grrenhouse gas intensity	P4 Non-renewable materials intensity
	O5 Residuals intensity	P5 Restricted substances content
	O6 Air relases intensity	P6 Energy consumption intensity
	O7 Water releases intensity	P7 Greenhouse gas emission intensity
	O8 Proportion of naturel land	

Source OECD (2011)

the public sector), and environmental performance (indicators such as resource use, waste management, and health risks to current and future generations) (Nikolaeva and Bicho 2011).

The inclusion of multiple aspects of sustainability performance in the GRI framework has allowed companies to manage their performance in a consistent manner and to compare it to other countries (Azapagic and Perdan 2000). Similarly, they explain that external stakeholders (especially financial ones) can use benchmarking to compare companies on the basis of their performance. As a result, and in particular, environmental performance indicators face many expectations (Tyteca et al. 2002). There is, however, one overriding objective: to improve the transparency—both internal and external—of environmental impacts and responses to mitigate them (Tyteca et al. 2002). It is therefore important to harmonize indicators internally and align them with internationally recognized indicators, in order to be able to position and define sustainable strategies.

The presentation of GRI indicators is always structured on two levels: core and supplementary indicators, regardless of the area of application (economic, environmental, social, employment practices and decent work, human rights, social, and product responsibility) (GRI 2015). According to the same source the "Environment" category covers impacts related to inputs (e.g. energy and water) and outputs (e.g. emissions, effluents and waste). It also covers impacts related to biodiversity, transportation, products and services, as well as environmental compliance and expenditure. Table 7.6 describes some indicators for the "Environment" category.

Table 7.6 Indicateurs de performances environnementales selon GRI G4

Element	Example of indicators
Materials	EN1—Material consumption by weight or volume EN2—Percentage of materials consumed from recycled materials
Energy	EN3—Energy consumption within the organization EN4—Energy consumption outside the organization EN5—Energy intensity EN6—Reduction of energy consumption EN7—Reduction of energy needs of products and services
Water	EN8—Total volume of water withdrawn by source EN9—Water supplies heavily impacted by withdrawals EN10—Percentage and total volume of water recycled and reused
Biodiversity	EN11—Operational sites owned, leased or managed in or adjacent to protected areas, as well as in biodiversity-rich areas outside protected areas EN12—Description of the substantial impacts of activities, products and services on the biodiversity of protected areas or biodiversity-rich areas outside protected areas EN13—Habitats protected or restored EN14—Total number of IUCN Red List and national equivalent threatened species whose habitats are in areas affected by activities, by level of extinction risk
Emissions	EN15—Direct greenhouse gas emissions EN16—Indirect energy-related greenhouse gas emissions EN17—Other indirect greenhouse gas emissions EN18—Greenhouse gas emissions intensity EN19—Reduction of greenhouse gas emissions EN20—Emissions of ozone-depleting substances EN21—NO_x, SO_x, and other substantial air emissions
Effluents and waste	EN22—Total water discharges, by type and destination EN23—Total weight of waste by type and treatment method EN24—Total number and volume of substantial spills EN25—Weight of waste transported, imported, exported, or treated and deemed hazardous (according to the Basel Convention), and percentage of waste exported EN26—Identification, size, protection status, and biodiversity value of water bodies and their ecosystems significantly affected by the organization's discharge and runoff

(continued)

Table 7.6 (continued)

Element	Example of indicators
Products and services	EN27—Extent of mitigation of environmental impacts of products and services EN28—Percentage of products sold and their packaging recycled or reused, by category
Compliance	EN29—Number of substantial fines and total number of non-monetary sanctions for non-compliance with environmental laws and regulations
Transport	EN30—Substantial environmental impacts from the transportation of products, other goods and materials used by the organization in the course of its business, and from the travel of its staff members
Generality	EN31—Total environmental protection expenditures and investments
Environmental assessment of suppliers	EN32—Percentage of new suppliers screened using environmental criteria EN33—Actual and potential significant adverse environmental impacts in the supply chain and actions taken

The value of GRI guidelines lies in the availability of information and explanations accompanying each indicator, facilitating understanding and integration through a well-described procedure and data sources to calculate the indicator.

7.4.2 Selection of Indicators Performance According to ISO 14031

In addition to social reporting tools, the 14000 series of standards have allowed for the effective and efficient development of environmental performance evaluation (EPE) systems. According to the standard (ISO 14031 2013), EPE is a continuous process of collecting and evaluating data and information to enable ongoing performance evaluation and monitoring over time. It provides industry with a standardized, comprehensive and detailed management tool that meets the expectations of both the organization itself and its stakeholders. Some experts claim that ISO 14031 is the best standard for environmental management systems, because it is based on the "Pan-Do-Check-Act" approach, and focuses directly on environmental protection, cleaner production, sustainable development and communication of the corresponding achievements, without imposing written procedures and system documentation on companies, as required by ISO 14001 (Jasch 2000). It is therefore particularly relevant to small and medium-sized enterprises (Jasch 2000).

The ISO 14031 standard distinguishes two categories of indicators: (1) Environmental State Indicators, (2) Environmental Performance Indicators:

1. **Indicators Environmental Status (IESs)** provide information on the relationship between the state of the environment and the organization's activities. They generally reflect the environmental impact or modification of the environment by industrial activities (e.g. increase in earth temperature, modification of aquatic fauna and flora, etc.).

   However, it is necessary to realize that an organism is not responsible for the environmental state alone. Therefore, it will not be surprising to find that despite the efforts made by organizations in terms of prevention and environmental protection, the effects on the natural environment are not very visible (Gallez and Moroncini 2003).
2. **Environmental performance indicators** must help measure the efforts of the organization and assess the impact of these efforts on operations (ADEME 2008). They are thus divided into two types:

   - **Management Performance Indicators (MPI):** also known as "system indicators". They translate information about management's efforts to develop environmental practices into actual environmental performance.
   - **Operational Performance Indicators (OPI):** provide information on the environmental performance of the organization's operations. They are directly linked to environmental aspects represented by flows of materials and energy exchanged with the environment (greenhouse gas emissions, discharge of organic pollutants into water, etc.).

Table 7.7 lists some examples of indicators for the different categories mentioned above.

It is important to note the diversity and multitude of indicators proposed, whether in the framework of GRI or ISO 14031. As a result, the process of selecting the most appropriate indicators is very important. It will depend on the production process itself, the company's strategy and objectives, the stakeholders and the steering needs

Table 7.7 Example of performance indicators according to 14031

The indicators of the Environmental Status
Air (Concentration of a specific pollutant in ambient air, measured at specific monitoring points, measurement of GHG reduction or removal...)
Water (Concentration of a specific pollutant in groundwater or surface water, dissolved oxygen in receiving waters, changes in groundwater levels...)
Soil (Concentration of specific nutrients in soils adjacent to the organization's facilities, protected areas in a particular local area, concentration of a specific pollutant in surface soils at given points in the area surrounding the organization's facilities...)
Flora (The evolution over time of crop yields in nearby fields, the total number of plant species identified in a particular local area...)
Fauna (Specific measures of habitat quality of individual species at the local level, the concentration of a specific pollutant in the tissues of a particular animal species present in the local or regional area)
Human beings (weighted average of noise pollution levels, the incidence of specific diseases, especially on sensitive populations, as shown by epidemiological studies conducted in the local or regional area)
Aesthetics, heritage and culture (condition of sensitive buildings, condition of spaces considered sacred in the vicinity of the organization's facilities...)
Environmental performance indicators

(continued)

Table 7.7 (continued)

Management Performance Indicators (MPI)	*Operational Performance Indicators (OPI)*
Related to management policies and programs: • *Number of objectives achieved* • *Percentage of employees trained vs. percentage of employees requiring training …* **Related to compliance and regulations:** • *Degree of regulatory compliance* • *Time required to respond to environmental incidents* • *Percentage of successful and unsuccessful corrective actions* • *Frequency of emergency drills conducted …* **Related to financial performance:** • *The costs (operating and capital) associated with the environmental aspects of a product or process* • *The return on investment of environmental improvement projects* • *The savings achieved through the reduction of resources used, pollution prevention and waste recycling …* **Relating to community relations:** • *The number of press reviews regarding the organization's environmental performance* • *Progress of local remediation activities (e.g., local clean-up or recycling initiatives …*	**Materials:** • *The amount of material used per unit of product* • *The quantity of raw materials reused in the production process* • *The quantity of toxic materials used in the production process …* **Energy:** • *The amount of energy used per year or per unit of product* • *The quantity of each type of energy used (e.g. renewable)* • *The quantity of energy units saved through energy saving programs …* **Services useful to the organization's operations:** • *The amount of hazardous cleaning agents used by service providers* • *The type of waste generated by contracted service providers …* **Physical facilities and equipment:** • *The annual number of emergency events (e.g., explosions) or non-routine operations (e.g., shutdowns)* • *The number of annual operating hours of a particular piece of equipment …* **Supplies and deliveries:** • *The number of deliveries for each mode of transportation per unit time* • *The percentage of business meetings conducted remotely* • *Percentage of fleet vehicles equipped with pollution-reducing technology …* **Products:** • *Number of products that can be reused or recycled* • *Percentage of product content that can be reused or recycled* • *Percentage of defective products* • *Percentage of products with instructions for environmentally safe use and disposal …* **Services provided by the organization:** • *Resource consumption per unit of service provided* • *Quantity of pollutants per unit of service provided* **Waste:** • *Amount of waste per unit (e.g., product, time, labor)* • *Quantity of hazardous, recyclable or reusable waste generated per unit* • *Total amount of waste disposed of by category* • *Quantity of waste disposed as a result of pollution prevention programs* **Emissions:** • *Amount of specific emissions per year* • *Quantity of emissions to air that may cause ozone depletion* • *Quantity of specific materials discharged to water per unit of product* • *Noise pollution measured in a significant location*

of the recipient of these indicators. It is possible that for the same environmental aspect several indicators can be associated with it depending on the recipients and uses (ISO 14031 2013). Consequently, the choice of indicators must seek coherence with the environmental policy and establish a continuity that goes from environmental aspects to performance indicators (ADEME 2008), including:

- With regard to the credibility requirements of internal and external stakeholders, they must therefore be relevant and understandable from their point of view,
- Depending on their representativeness of environmental performance, their sensitivity to changes in environmental performance,
- Depending on their availability and cost.

It should be remembered that the choice of environmental performance indicators can be greatly facilitated by regulations. Indeed, in several countries, environmental criteria must be respected. Therefore, the company, in its search for compliance with the regulations and consequently with ISO 14001, can establish indicators, for example, measuring the emissions of a specific pollutant or the discharge of an effluent subject to legal texts. The procedure for identifying indicators according to the OECD is described in Table 7.8.

7.4.3 Environmental Performance Assessment Methods

The author (Personne 1998) distinguishes four main categories of environmental performance evaluation methods with different structures and objectives:

- **Diagnostis methods**: based essentially on multiple choice questionnaires. The objective of these methods is to raise the company's awareness of the environmental issue by detecting approximately the areas (water, energy, waste, liquid discharges, etc.) where there are weak points. The advantage of these methods lies in their simplicity of use, which does not require great technical skill. Therefore, they are perfectly adapted to SMEs that have little or no in-house environmental skills. This type of assessment is not sufficient, but the company can become aware of the extent of the commitment to environmental practices.
- **Initial assessment methods:** These offer an exhaustive material/energy assessment that leads to the construction of an initial corrective action plan. Unlike the first category, this one is aimed at companies with internal environmental skills. The assessment is based on raw data from the field and offers a methodological guide for the construction of an action plan from the study and understanding of the production system and the identification of sources of pollution.
- **EMS Audit Methods**: Verify the proper functioning and compliance of the company's EMS with the requirements of the ISO 14001 standard. Carried out periodically and cyclically, the EMS audit provides feedback after the implementation of the program. This does not prevent the audit method from being used as an initial assessment to take stock of the situation. In addition, the auditor may be

Table 7.8 Seven steps to utilize the OECD sustainable manufacturing indicators

	Step	Comment
Prepare	1. Map your impact and set priorities	Learn how to bring together an internal "sustainability team" to set objectives, review your environmental impact and decide on priorities
	2. Select useful performance indicators	Identify indicators that are important for your business and learn about what data should be collected to help drive continuous improvement
Measure	3. Measure the inputs used in production	Identify how materials and components used into your production processes influence environmental performance
	4. Assess operations of your facility	Consider the impact and efficiency of the operations in your facility (e.g. energy intensity, greenhouse gas generation, emissions to air and water)
Improve	5. Evaluate your products	Identify factors such as energy consumption in use, recyclability and use of hazardous substances that help determine how sustainable your
	6. Understand measured results	Learn to read and interpret your indicators and understand trends in your performance
	7. Take action to improve performance	Choose opportunities to improve your performance and create action plans to implement them

Source OECD (2011)

internal or external. For certification, the external audit is necessarily conducted by an independent third party, whereas the internal audit is more successful in terms of improving the EMS. The application of this type of method assumes that the company is already engaged in the implementation of an EMS.

- **Multi-phase methods**: Assist companies to progressively address environmental issues through several evaluation phases: (1) awareness raising, identification of environmental constraints and challenges for the company; (2) identification of environmental impacts due to the company's activity; (3) evaluation of environmental impacts; (4) possible proposal of improvement solutions (technical or organizational) and implementation; (5) monitoring of environmental performance for continuous improvement.

Other authors such as Hermann et al. (2007) outline some of the tools and methods that are being used in industry to gather detailed information on environmental aspects and impacts. The most powerful methods recognized so far are life cycle assessment (LCA), multi-criteria analysis (MCA) and environmental performance indicators (EPI).

From Table 7.9, it can be seen that some approaches are product-centric (LCA) and others are more system-centric (EPI's). But there are other more specific approaches that are cited in the literature such as: ecological footprint, Eco-efficiency, Material flow analysis (Cikankowitz 2008). According to the same author, there are also approaches that focus on a theme or type of impact such as waste, energy or air emissions (e.g., carbon footprint).

As for the environmental performance indicator (EPI) approach, it is a classic approach based on measuring and then comparing the current and past results of

Table 7.9 Comparison of environmental assessment methods according to Hermann et al. (2007)

	LCA	MCA	EPIs
Purpose of the analysis	To compile and evaluate the environmental impacts of a product over its entire life cycle	To evaluate the overall environmental consequences of an alternative, taking into account multiple criteria and their relative weights	To compare an organisation's past or present environmental performance to its environmental performance criteria
Procedure	Goal and scope definition, inventory analysis, impact assessment and interpretation	Establishing the decision context, identifying criteria, scoring, weighting deriving an overall value, examining the results and conducting a sensitivity analysis	Data collection, establishment of a database, aggregation, standardization or normalisation into a set of performance indicators
Final output of the instrument	A limited set of environmental scores for a number of impact categories	One environmental score based on an aggregation of criteria	Environmental scores for a large number of performance indicators (consumption of raw materials etc.)
Strengths of the instrument	Avoids problem shifting to other issues or areas, comprehensiveness through 'cradle-to grave' approach	Possibility of weighting the criteria, use of criteria with their own dimensions, single score for overall evaluation	Can be designed individually for each organisation, limited time and data requirements, widely used for benchmarking
Weaknesses of the instrument	LCA is a complex process and requires considerable time and data input; dependence of normalisation on reference scenario; difficulties in interpreting the results	MCA usually only takes a part of the production chain into account; relies on input from experts and stakeholders; weighting is subjective	Indicators are only built for aspects on which data is readily available; large sets of indicators; lack of impact calculation (i.e. consequences of emissions); difficult to obtain the overall environmental performance

an organization with the objectives set by its management (Jasch 2000). It is also a method for reporting and communicating on the environmental performance of the company. In this sense, it also offers the possibility of benchmarking within the same sector (Azapagic and Perdan 2000; Tyteca et al. 2002).

Despite the difference in terms of advantages and disadvantages of the above mentioned methods, they are still complementary and should be used simultaneously. For example, LCA remains strongly linked to the IPE approach since it provides essential information on the potential impacts of the product on the environment during its entire life cycle. Since the phases of an LCA are: definition of objectives, inventory (material-energy balance over the entire life cycle), analysis of impacts, comparative evaluation of environmental impacts (Personne 1998). Moreover, the standard (ISO 14044 2006) specifies that LCA can help in the selection of relevant environmental performance indicators, including measurement techniques. For these reasons, ISO 14001 recommends the identification of environmental aspects and impacts from a life cycle assessment perspective.

7.5 Social and Societal Performance

According to our perception, the measurement of social and societal performance is directly linked to the actions undertaken by an organization in the framework of CSR with respect to its impacts on society and the environment. However, in what follows, only the "society" component will be taken into consideration insofar as the term "social" concerns the conditions that make it possible to make the contributions and rewards of employees equitable; and the term "societal" concerns the relationship with the communities in the vicinity of the company (Yedder and Farhoud 2009). In this sense, it is often admitted that taking this concept into account in the company's strategy implies the presumed existence of an implicit contract between the company and society, according to which the company has obligations towards society, which has the right to control it. This is particularly true with the advent of GRI, which has become a benchmark for "voluntary" social reporting throughout the world, and which instrumentalizes external communication on social and societal performance. Thus, the measurement of this type of performance is intimately linked to taking into account the requirements and expectations of a broader set of stakeholders. As stated by Saulquin and Schier (2007) "To develop the company on a sustainable and profitable basis by balancing the aspirations of customers, work teams, suppliers and shareholders in order to contribute positively to the society in which we live". This was also reported by Clarkson (1995) who defines societal performance as "the ability to manage and satisfy stakeholders". But can this notion of satisfaction alone be used to evaluate societal performance? Obviously, the assessment of this type of performance is very delicate and requires the following aspects to be taken into account (Martory 2003):

- **The means implemented:** in the absence of results indicators, it is recommended to measure, at least, the efforts made in the social (internal) and societal (external) fields. For the company is evaluated according to its social commitments put into practice through actions that promote respect for fundamental human rights at work and the collective interests of the society in which it is located (Forman et al. 2007).
- **The results obtained**: according to the degree of achievement of the objectives pursued (effectiveness).
- **Assessment of performance**: relating the results obtained to the means implemented in a logic of efficiency (return on investment in training, for example).
- **Risks:** a social or societal risk is generally defined as a potential loss of value (e.g. penalties for non-compliance with standards, regulations, conflicts with employees or consumers, etc.).

As for social performance, it focuses on the human aspect in organizations (Bernard 2009). According to the same source, it is linked to the results of human resources management in relation to criteria of staffing, operating mode structures, employee satisfaction and mobilization.

7.5.1 *Indicators and Evaluation Fields*

The notion of social and societal performance highlights the complexity of considering the multiplicity of expectations and interests of different internal and external stakeholders. Measuring this performance by integrating variable objectives (individual and collective, short and long term…) is a challenge for any company. To guide managers in their quest, eight dimensions are often cited in the literature as being taken into account by companies when evaluating social and societal performance (Jahmane and Louart 2013):

- **Business ethics:** contractual arrangements, fair practices with suppliers and customers, company policy in selecting suppliers, etc.
- **Local initiative:** involvement in local communities, strengthening relationships and communication, contributing to the local economy, striving for local enrichment and capacity building, etc.
- **Environment:** protection of natural resources, human health, etc.
- **Governance:** democracy, transparency, anti-corruption, etc.
- **Human rights:** child labor, forced labor, discrimination, freedom of association
- **Workplace:** wages and benefits, education and training, working hours, health and safety, job insecurity, etc.
- **Corporate involvement:** donations, grant and sponsorship programs, etc.
- **Social impact of products:** contribution of products and services to social welfare and equity.

Also, given the variety of stakeholders and the multiplicity of approaches, and in order to delimit the fields of evaluation, reference is often made to a recognized framework. Such as the GRI for social reporting, which is essentially a voluntary approach by the company. It consists of a report that takes into account its guidelines and is divided into five parts (ADEME 2008): (1) Corporate vision and strategy, (2) Presentation of the company, (3) Structure of governance and management systems, (4) An index specifying where the requested information is presented in the report, (5) Performance indicators broken down by component (economic, social, environmental). Indeed, GRI has contributed to the creation of a common language that can be used by corporate stakeholders to form an opinion on reported performance (Nikolaeva and Bicho 2011).

For example, to measure social performance, GRI (2015) proposes four Social Dimension Subcategories: (1) Employment Practices and Decent Work; (2) Human Rights; (3) Society; and (4) Product Responsibility. Most of the content of the Sub-Categories is based on internationally recognized universal standards or other relevant international references. Table 7.10 groups the indicators related to the "Employment Practices and Decent Work" Sub-Category.

It should be remembered that the commitment to a CSR approach takes into consideration the three classic dimensions already recommended by Elkington (1997) but two other dimensions can be added explicitly "Stakeholders" and "Volunteering" (Dahlsrud 2008). The interest of this categorization makes it possible to highlight the fundamental elements of CSR through stakeholders (social and societal) and volunteerism. This last dimension emphasizes the fact that CSR is essentially conducted through actions (not mandatory or regulated by law) of cooperation with the local community.

In addition, despite the importance of the evaluation and measurement of social and societal performance in the development of companies, it has been found in the literature that human and social factors have long been under, estimated although they are key parameters (Tahon 2003). This was also proven via a study conducted by Tarquinio et al. (2018) whose results highlighted a low disclosure of indicators related to human resources. On the other hand, they specify that social indicators related to work are those that are most frequently returned in sustainability reports.

It is noted that the measurement of this type of performance, for those who practice it, remains today governed mainly by objectives and purposes of external communication on CSR (Germain and Gates 2007). This voluntary communication and the adoption of the GRI would help companies to obtain a certain legitimacy as responsible corporate citizens by demonstrating their adherence to these benchmarks (Nikolaeva and Bicho 2011). In this sense, Saulquin and Schier (2007) states that CSR is an opportunistic performance lever, it involves communication strategies and a change in stakeholder relationships.

A study by Germain and Gates (2007), whose objective was to observe the place of CSR-related performance indicators in Sustainable Balanced ScoreCard-type management tools, showed that they are on average relatively absent from the management tools of the companies studied. This observation differs according to the organization: industrial companies that are sensitive to environmental risk and

Table 7.10 Social category indicators, subcategory: labour practices and decent work

Aspect	Indicateurs
Employment	LA1—Total number and rates of new employee hires and employee turnover by age group, gender and region LA2—Benefits provided to full-time employees that are not provided to temporary or part-time employees, by significant locations of operation LA3—Rate of retention and return after paternity leave by gender
Labour/ management relations	LA4—Minimum notice periods regarding operational changes, including whether these are specified in collective agreements
Occupational health and safety	LA5—Percentage of total workforce represented in formal joint management-worker health and safety committees that help monitor and advise on occupational health and safety programs LA6—Type of injury and the rates of injuries, occupational diseases, lost days, absenteeism and total number of work-related fatalities, by region and gender LA7—Workers with high incidence or high risk of illness related with their occupation LA8—Health and safety topics covered by formal agreements with trade unions
Training and education	LA9—Average number of hours of training by year, by employee, gender and by category LA10—Programs for Skills management and lifelong learning that support the continued employability of employees and assist them in their careers management LA11—Percentage of employees who regularly receive performance and career development analyses, by gender and professional category
Diversity and equal opportunity	LA12—Composition of governance bodies and breakdown of employees per employee category according to gender, age group, minority group membership, and other indicators of diversity
Equal pay for women and men	LA13—Ratio of base salary of men in comparison with that of women, by professional category and significant operating units
Supplier assessment for labour practices	LA14—Percentage of new suppliers who were selected using the criterion of labour practices LA15—The main real and potential impacts of labour practices on the supply chain and actions taken
Labour practices grievance mechanisms	LA16—Number of grievances about labour practice filed, addressed and resolved through formal grievance mechanisms

Source GRI (2015)

companies listed on the stock exchange that are subject to shareholder pressure are increasingly concerned about CSR issues and tend to develop strategic CSR management more than other companies.

7.5.2 Assessment Procedures

It is known that the commitment to social responsibility can be translated into practices linked to standards such as ISO 26000 or Social Reporting, which nevertheless provide a framework for evaluation. However, the fact that the ISO 26000 standard is not "certifiable" makes the question of the evaluation process perplexing. As a result, companies are wondering how to demonstrate their approach and commitment to social responsibility. Also, the lack of available evaluation guidelines leaves companies confused as to the choice of relevant indicators to adopt.

Therefore, companies are looking for other ways to value their actions, either through self-assessment or the evaluation of their performance. Thus, companies have turned to models of excellence and other existing instruments to promote their CSR approach (Table 7.11), knowing that they differ in terms of objectives and the means they propose.

The choice of an assessment instrument remains dependent on the objective pursued by the company (Reporting, Evaluation or Self-assessment). According to Jankalová and Jankal (2017), the most frequently cited and studied excellence models with a relationship to social responsibility are EFQM and MBNQA. Although they

Table 7.11 CSR models, standards, guidelines and indexes a comparison by Avlonas (2004, in Jankalová and Jankal 2017)

CSR approach	Tool for reporting	Tool for self-assessment	Tool for assessment
The EFQM Excellence Model	++	+++	+++
Social Accountability—SA 8000	+	+	+++
ISO 14OOO		++	+++
EMAS	+	++	+++
AccountAbility 1000 (AA1000)	++	+	+++
Global Reporting Initiative (GRI)	+++	++	
Value Management System (VMS)		++	++
Dow Jones Sustainability Index	+	++	
FTSE4good		++	

are intended and designed initially for a TQM (Total Quality Management) approach, these models directly propose performance measures in relation to stakeholders such as: customer, employee, society, partner and supplier perception results. They offer a practical framework that is recognized worldwide and can lead to external recognition, however, they are also vectors of internal transformation through the structuring and evaluation of sustainable performance. As Nikolaeva and Bicho (2011) point out, voluntary reporting according to GRI can be a resource-intensive, but also a transformative experience.

Moreover, these models focus not only on direct outcomes but also on practices and means to achieve them. As a result, they provide organizations with a general framework for self-assessment, evaluation and reporting on CSR. Moreover, they are compatible with all types of organizations and sectors of activity.

Also, the study (Jankalová and Jankal 2017) states that EFQM in Europe and MBNQA in the USA have recently become, with the increase in the importance of corporate social responsibility, the most appropriate assessment frameworks for CSR activities. They state that the CSR concept is indeed present in an integrated way in these models of excellence, the difference lies in the rate of coverage of individual CSR elements in each of its models (Jankal and Jankalova 2016).

Similarly, there are other references for assessing social and societal performance such as the AFAQ 1000NR model (Aubrun, et al. 2010). Similar to these counterparts, it has the advantage of focusing not only on results but also on strategic and managerial components. Indeed, the evaluation model is composed of two parts: the first part relates to the evaluation of the strategic and managerial practices of the company, the second focuses on the results monitored and achieved by the company in environmental, social and economic matters.

Thus, the AFAQ 1000NR evaluation analyzes the results in light of the practices carried out in the company, thus making it possible to evaluate the degree of maturity of the CSR/Sustainable Development approach within the company (Aubrun et al. 2010). According to the same authors, this model consists of five practice criteria (see Table 7.12) and three result areas (Table 7.13).

The results are expressed through the "pillars of sustainable development". It is noted that the composition in terms of indicators is in perfect harmony with those proposed by GRI.

Still within the framework of social performance measurement, which is carried out through objectives and indicators directly linked to the company's human capital, the authors (Huselid et al. 2005) propose a social scorecard following the example of the Blanced ScoreCard. It includes four indicators: (1) Employee and management behavior, (2) Employee success, (3) Employee mindset and culture, and (4) Employee skills.

It is noted that the measurement of social and societal performance reveals a great deal of complexity and diversity in assessment instruments. This is perhaps due to its direct attachment to the concept of CSR and the difficulty of dissociating them. However, the scientific and practical communities are showing their concern and great interest in proposing a tool that is accessible in terms of understanding and deployment.

Table 7.12 Five practice criteria AFAQ 1000NR

1. Vision and Strategy	• Management leadership and commitment • Openness to stakeholders • Identification of environmental issues • Identification of social issues • Identification of economic issues • Prioritization of issues • Decision-making process • Strategy/policy/objectives • Benchmark • Consistency with international benchmarks
2. Management of the process	• Identification of responsibilities • Relevance and efficiency of processes • Technological and competitive intelligence • Regulatory monitoring • Information and data management • Human resources and strategy • Human resources: knowledge and skills • Internal communication • External communication
3. Mode of production and consumption, and sustainability of products/services	• Management of purchasing and supplier relations • Production management and social aspects • Production and environment • Design • Global reflection on the activity
4. Territorial anchoring	• Territorial integration and risk analysis • Environmental development of the territory • Social development of the territory • Economic development of the territory • Transportation and accessibility of the site
5. Human Resources Management	• Management of social relations • Involvement and accountability • Motivation • Human resources management consistent with the principles of sustainable development

7.6 Conclusion

Sustainable performance is by nature "multidimensional", shaped by the culture of the organization, combining economic, operational, social, societal and environmental aspects, often complementary but necessarily converging towards the company's strategic axes. However, despite the divergence and the evolving nature of its definition, sustainable performance remains a construct specific to each company as long as it depends on a reference linked to the strategies defined and objectives set.

The evaluation of this "Global" performance is carried out through indicators and measurement systems such as the Sustainable Balanced Scorecard. It is a combination

Table 7.13 The three outcome domains according to the AFAQ 1000NR assessment model

Theme	Kpi example
Environmental results	
Raw material	Percentage of raw materials from recycling channels
Energy	Share of renewable energy in total consumption
Water	Type of water supply sources
Biodiversity	Habitat area protected or restored
Greenhouse gases	Emissions monitoring
Emissions, effluents and waste	Quantity of waste by category and destination
Products and services	Percentage of recovered and eco-designed products
Compliance with agreements and regulations	Membership in the Golobal Compact
Social results	
Employment	Percentage of local labor
Labor Relations	Monitoring of consultation rules
Health and safety	Monitoring of International Labour Organization indicators
Training and education	Number of training hours
Diversity (equal opportunities and anti-discrimination) and equity	Ratio of male to female employees
Human rights	Percentage of suppliers selected on such criteria
Civil society	Percentage of employees trained in anti-corruption practices
Product responsibility	Consumer accidents related to the use of products
Economic results	
Economic performance	Percentage of contracts honored within agreed terms
Market presence	Share of local suppliers
Pricing policy	Cost of all goods
Investments (in clean technologies, environment…)	Share of investment directly related to sustainable development projects

Source Aubrun et al. (2010)

of a "measurement tool" and a "management tool" (Saulquin and Schier 2007). In this sense, it constitutes for management and employees, a means of steering and assisting in decision-making at the level of the company and its processes, and thus provides the necessary information to make decisions and initiate actions. However, performance evaluation is not an end in itself but a means of enabling the company to be part of a continuous improvement process based on measurement.

Its benefits can also be appreciated externally, as long as it constitutes a means of communication on the level and impact of the company's efforts to achieve collective results. Indeed, this is rather valid for companies that practice social and environmental reporting and that have to communicate their results to stakeholders (state, investors, local authorities…).

However, the construction of a measurement system remains a very perilous task because of the multitude of choices of measurement axes (operational, environmental, and social), and the corresponding performance indicators. These indicators must reconcile the short and the long term, the qualitative and the quantitative, the local and the global, the collective and the individual interest… These different indicators must also provide sufficient information on the company's activity or its processes while focusing on its strategic objectives. This exercise requires both transversal management (harmonization between processes in terms of measurement axes) and vertical management (harmonization between processes and strategic axes).

Therefore, the implementation of an appropriate management system that complies with international standards (GRI, ISO 14001, ISO 26000, EFQM, etc.) remains the surest way to acquire control over performance measurement and also a means of proving the validity of the approach to the various stakeholders. These standards represent a guide for companies to proceed with an approach based on the construction of a measurement system through the identification of key indicators within the framework of an elaborated strategy.

References

ADEME Od (2008) La Gestion Environnementale. Dunod, Paris

AFNOR (2000) Système de Management de la Qualité: Indicateurs et tableaux de bord. FD X 50-171

Åhlström P (1998) Sequences in the implementation of lean production. Eur Manag J 6(3):327–334

Aubrun M, Bermond F, Brun E, Cortot J-L, Delchet-Cochet K, Graffin O, Jounot A, Ponrouch A (2010) ISO 26000 Responsabilité Sociétale: Comprendre, déployer, évaluer. AFNOR, Paris

Azapagic A, Perdan S (2000) Indicators of sustainable development for industry: a general framework. Trans IChemE 78(4):243–261

Baret P (2006) L'évaluation contingente de la Performance Globale des Entreprises: une méthode pour fonder un management sociétalement responsable? In: Rosé DJ-J (ed) Responsabilité sociale de l'entreprise. Pour un nouveau contrat social. De Boeck Supérieur, pp 135–152

Belhadi A, Touriki F-E, El-Fezazi S (2018) Development of a lean assessment tool for small and medium sized-enterprises. In: Viles DE, Ormazabal M, Lleo A (eds) Closing the gap between practice and research in industrial engineering. Lecture Note in Management and Industrial Engineering. Springer, pp 361–369

Bernard M (2009) L'effet des pratiques de ressources humaines sur la performance sociale des employés dans un contexte de culture nationale. Université de Strasbourg: Ecole de Management Strasbourg, Strasbourg

Bieker T, Gminder CU (2002) Managing corporate social responsibility by using the "sustainability-balanced scorecard". In: Contribution to the 10th International Conference of the Greening of Industry Network. Göteborg (Sweden), p 17

Bouamama M (2015) Nouveaux défis du système de mesure de la performance: cas des tableaux de bord. Thèse, Université de Bordeaux

Brown HS, De Jong M, Lessidrenska T (2009) The rise of the Global Reporting Initiative: a case of institutional entrepreneurship. Environ Polit 18(2):182–200

Cikankowitz A (2008) Méthodologie d'évaluation des performances environnementales de techniques en vue de les comparer puis de les valider "meilleures techniques disponibles. Thèse, Ecole Nationale Supérieure des Mines, Saint-Etienne
Clarkson MB (1995) A stakeholder framework for analyzing and evaluating corporate social performance. Acad Manag Rev 20(1):92–117
Cocca P, Alberti M (2010) A framework to assess performance measurement systems in SMEs. Int J Product Perform Manag 59(2):186–200
Dahlsrud A (2008) How corporate social responsibility is defined: an analysis of 37 definitions. Corp Soc Responsib Environ Manag 15:1–13
Duque DF, Cadavid LR (2007) Lean manufacturing measurement: the relationship between lean activities and lean metrics. Estudios Gerenciales 23(105):69–83
EFQM (2021) The EFQM excellence model. Consulté le Mars 20, 2021, sur https://www.efqm.org/
Elkington J (1997) Cannibals with forks: the Triple bottom line of 21st century business. Capstone Publishing Ltd., Oxford
Fercoq A (2014) Contribution à la modélisation de l'intégration lean green appliquée au management des déchets pour une performance équilibrée (économique, environnementale, sociale). Thèse, Ecole Natioanale Supérieure d'Arts et Métiers-ENSAM, Paris
Figge F, Hahn T, Schaltegger S, Wagner M (2002) The sustainability balanced scorecard—linking sustainability management to business strategy. Bus Strat Environ 11:269–284
Forman B, Gey J-M, Bonnifet F (2007) Qualité Sécurité Environnement: Construire un système de management intégré. AFNOR
Gallez C, Moroncini A (2003) Le manager et l'environnement: outils d'aide à la décision stratégique et opérationnelle. Presses Plytechniques et Universitaires Romandes
Garengo P, Bernardi G (2007) Organizational capability in SMEs performance measurement as a key system in supporting company development. Int J Prod Perform Manag 56:518–532
Germain C, Gates S (2007) le niveau de developpement des indicateurs de responsabilite sociale dans les outils de pilotage de controle de gestion: une analyse des pratiques des entreprises. Comptabilité et Environnement, France, p 15
GRI (2015) G4 Lignes directrices pour le reporting développement durable. Global Reporting Initiative, Amsterdam
Hansen EG, Schaltegger S (2014) The sustainability balanced scorecard: a systematic review of architectures. J Bus Ethics 133:193–221
Hennion R, Makhlouf A (2016) Les fiches Outils du Lean Six Sigma. Eyrolles, Paris
Hermann B, Kroeze C, Jawjit W (2007) Assessing environmental performance by combining life cycle assessment, multi-criteria analysis and environmental performance indicators. J Clean Prod 15:1787–1796
Hollnagel E, Woods DD, Leveson NC (2006) Resilience engineering: concepts and precepts. Aldershot, Ashgate
Huselid M, Becker B, Beatty R (2005) Tableaux de Bord Sociaux: Mesurer la contribution des salariés à l'exécution de la stratégie. Pearson
ISO 14001 (2015) Environmental management systems-requirements with guidance for use
ISO 14031 (2013) Environmental management—environmental performance evaluation—guidelines
ISO 14044 (2006) Environnemental management-life cycle assessment—requirements and guidelines
ISO 9000 (2015) Système de management qualité-Principes essentiels et Vocabulaire
Jahmane A, Louart P (2013) La Responsabilité Sociale de l'Entreprise: une diversité des concepts, des enjeux multiples et imbriqués et diverses méthodes de mesure. Manag Sci Soc: 99–117
Jankal R, Jankalova M (2016) The application of The EFQM Excellence Model by the evaluation of Corporate Social Responsibility activities of companies. In: 3rd Global Conference on Business, Economics, Management and Tourism. Elsevier, Procedia Economics and Finance, Rome, Italy, pp 660–667

Jankalová M, Jankal R (2017) The assessment of corporate social responsibility: approaches analysis. Int J Entrep Sustain Issues 4:441–459

Jasch C (2000) Environmental performance evaluation and indicators. J Clean Prod 8:79–88

Kamble S, Gunasekaran A, Ghadge A, Raut R (2020) A performance measurement system for industry 4.0 enabled smart manufacturing system in SMMEs—a review and empirical investigation. Int J Prod Econ 229:41p.

Kaplan RS, Norton DP (1992) The balanced scorecard-measures that drive performance. Harvard Business Review, pp 71–79

Kaplan RS, Norton DP (1996) Using the balanced socrecard as a strategic management system. Harvard Business Review, 13p.

Kaplan RS, Norton DP (2001) Comment utiliser le tableau de bord prospectif? Editions d'Organisation, Paris

Karlsson C, Åhlström P (1996) Assessing changes towards lean production. Int J Oper Prod Manag 16(2):24–41

Le Réseau du CAD sur l'évaluation en matière de développement (2017) Evaluer la Coopération pour le développement: récapitulatif des normes et standards de référence, Deuxième édition. OCDE

Lorino P (2003) Méthodes et Pratiques de la Performance. Editions d'Organisation, Paris

Manetti G, Becatti L (2009) Assurance services for sustainability reports: standards and empirical evidence. J Bus Ethics 87:289–298

Martory B (2003) Audit social: Pratiques et Principes. Revue Française de gestion 29:235–246

Maurel C, Tensaout M (2014) Proposition d'un modèle de représentation et de mesure de la performance globale. Comptabilité Contrôle Audit 3:73–99

Morin E, Guindon M, Bouliane E (1996) Mesurer la performance de l'entreprise. Gestion 21(3):61–66

Nikolaeva R, Bicho M (2011) The role of institutional and reputational factors in the voluntary adoption of corporate social responsibility reporting standards. J Acad Mark Sci 39:136–157

OECD (2011, May) Sustainable manufacturing toolkit: seven steps to environnemental excellence. Récupéré sur OECD: www.oecd.org/innovation/green/toolkit

Pakdil F, Leonard KM (2014) Criteria for a lean organisation: development of a lean assessment tool. Int J Prod Res 52(15):4587–4607

Personne M (1998) Contribution à la méthodologie d'intégration de l'environnement dans les PME-PMI: Évaluation des performances environnementales. Thèse, L'institut National des Sciences Appliquées, Lyon

Quairel F (2005) La mesure de la performance sociétale à l'aune de l'instrumentation comptable et financière. In: LeRoy DF, Marchesnay M (eds) La Responsabilité Sociale de l'Entreprise. EMS, p 352

Robbins S, Decenzo D, Coulter M, Rüling C-C (2014) Management: l'essentiel des concepts et des pratiques, 9th ed. Nouveaux Horizons

Sánchez AM, Pérez MP (2001) Lean indicators and manufacturing strategies. Int J Oper Prod Manag 21(11):1433–1452

Saulpic O, Giraud F, Zarlowski P, Lorain M-A, Fourcade F, Morales J (2011) Les Fondamentaux du contrôle de gestion. Pearson Education France

Saulquin J-Y, Schier, G (2007) Responsabilité sociale des entreprises et performance: complémentarité ou substituabilité? La Revue des Sciences de Gestion, Direction et Gestion (223):57–65

Tahon C (2003) Evaluation des performances des systèmes de production. Hermes/Lavoisier

Tarquinio L, Raucci D, Benedet R (2018) An investigation of global reporting initiative performance indicators in corporate sustainability reports: Greek, Italian and Spanish evidence. Sustain 10(4):897

Tyteca D, Carlens J, Berkhout F, Hertin J, Wehrmeyer W, Wagner M (2002) Corporate environmental performance evaluation: evidence from the MEPI project. Bus Strat Environ 11:1–13

Yedder M-B, Farhoud M (2009) Le développement durable est-il bienvenu dans les organisations? Cas de l'implantation d'un Système de Management Environnemental en Tunisie. Développement durable & territoire

Zingales F, Hockerts K (2003) Balanced scorecard & sustainability: examples from literature and practice. In: Schaltegger DS, Dyllick T (eds) Nachhaltig managen mit der Balanced Scorecard. Konzept und Fallastudien. Gabler, Weisbaden, pp 151–166

Chapter 8
Sustainable, Lean and Resilient SMEs in the Age of COVID 19

Objectives

1. Provide an overview of the sustainable, lean and resilient strategies and the adoption challenges
2. Provide a framework for the integration of lean-resilient and sustainable paradigms in SMEs

8.1 Introduction

The spread of the Covid-19 pandemic has affected various sectors of the economy, including the micro, small and medium enterprises (SMEs) sector. Unfortunately, the limitation of economic activities to social restrictions causes the ability of SMEs to face turmoil to become greatly limited. The disruption of business activities will even reduce the business performance; therefore, the companies make layoffs and face the threat of bankruptcy, some of these have even gone bankrupt because they are unable to survive (Humas LPDB-KUMKM 2020). This pandemic attracted the scholar to investigate how the various existing and emerging management philosophies or paradigms help in avoiding the aforementioned challenges and achieve their objectives to survive and to remain competitive. There a wide range of paradigms and techniques available in literature and in practice such as Lean, Resilient and sustainable, these paradigms are seen as the most used and relevant. Each of these paradigms have their respective objectives, focus and importance. However, they serve for a common purpose, that is, to make SMEs competitive, efficient and effective. Lean focuses on non-value-added activities or waste elimination, while resilient practices work towards how Buisness operations respond to the unexpected event, sustainability concept urges for embracing the triple bottom line concepts, which has all three dimensions, that is, economic, social and environmental. The aim of this study is the implementation of a Lean, Resilient and Sustainable approach to help SMEs facing post COVID pandemic.

F. E. Touriki et al., *Sustainable Excellence in Small and Medium Sized Enterprises*, Industrial Ecology, https://doi.org/10.1007/978-981-19-0371-7_8

This paper refers to SMEs implementation of Lean, Resilient and sustainable (LRS) paradigms and contributes to the argumentation about LRS in the business and management field by answering the research questions on how to implement resilient lean and sustainable strategies to face COVID pandemic. The purpose of this chapter is to answer the question of how to implement sustainable, lean and resilient paradigms in the field of SME's business and management facing, and rising from business turbulence caused by the Covid-19 pandemic. Therefore, we analyze multiple case studies and outline the conceptual framework of SMEs to adopt LRS paradigms. The proposed framework configuration can enhance the conceptualization of SMEs paradigms comprehensively and contribute to the business and management literature.

8.2 Background

8.2.1 COVID-19 and SMEs

There is widespread consensus on the fact that SMEs form the backbone of the global economy as they contribute to 35 per cent of the GDP in developing countries and around 50 per cent in developed countries WTO (2020).

Due to their nature and size, Asgary and Ozdemir (2020) claimed that SMEs are always vulnerable to any economic downturn and crises. They lack according to Giannacouroua, Maroula, and Christopoulouc (2020), the management mechanisms, the manpower, the material and the financial resources to deal with such unexpected events. However, the same authors claimed that SMEs are considered important in promoting national economic and social development since they employ more than half of the labor force in the private sector and, especially in the European Union, SMEs account for over 99% of all enterprises (OECD 2009).

Over the first half of 2020, The World has been gripped by a pandemic. Qiu (2020) said the virus jwas identified as a new coronavirus (severe acute respiratory syndrome coronavirus 2, or SARS-CoV-2), and later named as Coronavirus Disease-19 or COVID-19. Given the rapid spread of COVID-19, countries across the World have adopted several public health measures intended to prevent its spread, including social distancing reported by Fong (2020). The pandemic does not only affect the global health condition but also impending the structure of global economic order. Covid-19 is jeopardizing the economic well-being of peoples and institutions as described by (Sneader and Singhal 2020), according to an early forecast of the International Monetary the global economy would contract by about 3% in 2020. Consequently, many economies are in the dawn of recession (OECD 2020). As stated by the International Trade Centre (2020), COVID-19 negatively affected supply and demand. The pandemic induced lockdowns in China, the European Union and the United States of America, commonly known as group of three countries had major impacts on products, imports and exports which affected both big and small companies globally.

Juergensen et al. (2020) referred to the pandemic as an external shock of unprecedented magnitude, affecting SMEs on the supply and demand. In fact, Zeidy (2020) found that PMEs are faced with a decrease in demand, supply chain disruptions, cancellation of export orders, raw material shortages and transportation disruptions. Sourcing from new suppliers or absorbing price increases is more challenrging for small firms with limited supply options and capital.

R. Munoz Morezo (2020) assumed that the current situation led to a dramatic and sudden loss of demand for supplies which in turn led to a loss in revenue thus severely affecting the ability of most enterprises to function because of liquidity shortages. He estimated that the impact is more hardly felt among SMEs due to the type of economic activity they conduct, which often involve face-to-face interaction (i.e. retail, accommodation, tourism, and others). On the supply side, SMEs have faced logistical issues due to the disruption of transportation and labor shortages (Juergensen et al. 2020). On the demand side, SMEs have seen their demand decline substantially due to lockdown measures, a drop in consumer confidence and the shutting down. The severity of these demand and supply shocks is likely to differ according to whether the frm is a stand-alone, a specialized-supplier or a knowledge-based SME (Juergensen et al. 2020).

In addition, a study led by United Nation Economic Commission for Africa on the impact of COVID-19 on SMEs in Africa showed that most SMEs are mainly concerned about the cash-flow outlook because of disruptions in supply and demand. Actually, the study pointed that SMEs in developing countries have limited access to alternative financing mainly because most of them are informal and some do not have assets that can be used as collateral for new credit lines with financial institutions. A study by Piette and Zachary (2015) shows that banks perceive SMEs as presenting a higher risk during crises, which in turn explains tighter credit conditions. This, however, seems to apply mainly to SMEs which were not part of the existing corporate clientele pre-crisis. Risk perceptions can be triggered by innovative moves. In this regard Lee, Sameen, and Cowling (2015) find that SMEs have more problems accessing finance during crises, SMEs have done relatively worse and are particularly likely to face absolute credit rationing.

Moreover, Sandberg (2020) mentioned that the majority of SMEs operate in manufacturing, hospitality services and tourism sectors, thus these sectors were the most affected by the measures imposed by government to curb the spread of the virus. Indeed, C. Van Wyk (2020) argue these measures that were imposed worldwide exposed SMEs in the tourism and hospitality sector to the threat of closure thus rendering a significant number of people jobless. According to The International Labour Organisation, The lockdowns and quarantine measures caused a reduction of labour as most businesses which were regarded as non-essential were forced to close by governments. MSMEs are generally labour intensive and the COVID-related restrictions on the movement of workers negatively affected the day to day operations of these enterprises.

This pandemic represents an external shock of unprecedented magnitude, afecting European SMEs on the supply and demand sides alike Survey data from May 2020 suggest that 41% of UK SMEs had stopped operations and 35% feared they would be unable to reopen again (FSB 2020). In Germany, 50% of SMEs expected a negative efect due to the crisis with one-third anticipating a decline in revenues by more than 10% (DIHK 2020). In Italy, more than 70% indicated they were directly afected by the crisis (CNA 2020). While SMEs in other European countries have voiced similar concerns (OECD 2020). In Morocco, according to the survey, carried out by the HCP in April 2020 with businesses, 72% and 26% respectively shutdown production units of activity temporarily or permanently in April were very small businesses and SMEs. According to the survey of All India Manufacturer's Organization (AIMO) conducted for 5000, MSMEs said that 71% of the businesses were not ready to pay salaries in March 2020. That means financial crunch can be seen during these days. Production has stopped due to the non-availability of materials and workers. A survey conducted in China by researchers from Tsinghua University of 995 SMEs in February showed that 30% of the companies have seen their income drop by more than 50%; another 28% reported a 20 to 50% drop. More than one-third of the companies in the study reported that they could stay open for only one month with their current cash flow.

8.2.2 Sustainable, Lean and Resilient Strategies to Enhance SMEs' Business Excellence During COVID-19

According to Aisyah et al. (2019) several approaches have been taken to improve SMEs competitiveness. A lean approach, oriented towards reducing the cost of process improvement through elimination of waste and processes that have no added value. Resilient, the ability of the system to return to its original state after experiencing interference and avoiding a failure mode. Sustainability related to environmental, green industry management to reduce risks and environmental impacts while increasing ecological efficiency of organizations and partners.

Lean management approach, developed by Ohno (1988) at Toyota Motor Corporation in Japan, forms the basis of the Toyota Production System with two main pillars, autonomation and just-in-time (JIT) production. The focus of lean paradigm has essentially been on the waste reduction used for increasing actual value adding activities, to fulfill customer needs and maintaining profits. Reichhart and Holweg (2007) had extended the concept of lean production to the downstream or distribution level, lean distribution is defined as minimizing waste in the down-stream supply chain, while making the right product available to the end customer at the right time and location.

Resilience has been seen as the ability of the organization to simply bounce back from a distinctive, discontinuousevent that creates vulnerability and requires an unusual response (Lengnick-Hall and Beck 2003), Wildavsky (1988) further characterizes resilience as the capacity to cope with unanticipated dangers after they have become manifest. The concept of resilience is closely related with the capability and ability of an element to return to a stable state after a disruption. Van der Vegt et al. (2015) as well as Sterling (2011) show that the foundation of a company's resilience can be a the capacity to identify, prevent and manage risks.

Sustainable management defined by Seuring and Müller (2008) as the management of material, information and capital flows, as well as cooperation among companies along the supply chain, while taking goals from all three dimensions of sustainable development (economic, environmental and social), which contribute to sustainable development and are collectively referred to as triple bot-tom line (TBL). TBL is regarded as the base for successful sustainable development, and represents a metric for sustainability performance. For Jenkins (2009), sustainability refers to the ultimate goal of maintaining the viability of our economies, the society in which they operate and the environment on which they depend. From a managerial point of view, it is for the company to take into consideration the economic, social and environmental impacts of its activities Kechiche and Soparnot (2012). Table 8.1 discusses the Lean-Reslient-Sustainable paradigms, mostly based on the purpose, advantages and disadvantages.

The COVID-19 pandemic has had an uncertain impact on the global economy, especially for entrepreneurs and small and medium-sized enterprises that have suffered significant consequences. However, resilience has emerged as an entrepreneurial skill that allows companies to adapt and grow stronger in the face of challenges (Castro and Zermeno 2020). SMEs carry limited resources, so for the achievement of sustainability, SMEs need to take into account the environmental, economic and social issues through adopting lean and sustainable initiatives (Siegel

Table 8.1 The advantages and disadvantages of Lean-Reslient-Sustainable paradigms

	Lean	Resilient	Sustainable
Purpose	Waste elimination and valuecreation and apply concept ofzero inventory (Christopher 2000; Hines et al. 2004) or cos tminimisation through NVA elimination (Abdulmalek and Rajgopal 2007)	Recover the desired state with inpermissible time and cost, after experiencing unexpected	Higher competitive gain
Advantages	Reduction in set-up times provides internal manufacturing flexibility and enhances cost reduction possibility (Vonderembseet al. 2006)	Anticipate and act on the changes in the market and minimize the demand risk	By pursuing sustainability goal, SMEs differentiate themselves from the market
Disadvantages	Lack of external responsiveness to customers' demands (Vonderembse et al. 2006)	Have to keep strategic capacity and inventory buffer that is responsible for extra investment and cost	Expensive practices and product cost

et al. 2019), as both lean and green enhance SMEs competitiveness in a sustainable way (De et al. 2018). Therefore, SMEs are struggling to effectively integrate lean with green management (Farias et al. 2019).

Carvalho et al. (2011) provide a useful comparison of lean and resilient paradigms. According to these authors, each paradigm follows the following purpose: First Lean—looks for cost reduction and flexibility, for already available products, through continuous elimination of waste or non-value added activities across the chain. Second resilient is related to the system's ability to return to it is original state or to a new, more desirable, after experiencing a disturbance, and avoiding the occurrence of failure modes. While Lean paradigm requires low inventory level to reduce invent cost, the Resilient paradigm requires a strategic stock to respond effectively to unexpected disturbances along the chain. If the inventory of critical materials is maintained in low levels, the buisness will be more vulnerable to unexpected events that affect these materials supply (Carvalho et al. 2011).

Marchese et al. (2018) state that the relationship between resilience and sustainability can take three different forms: Sustainability and resilience as separate conceptual objectives; sustainability as a component of resilience; or resilience as a component of sustainability. The first form states that managing sustainability does not contribute to resilience, nor does managing resilience to sustainability. Its application and results are exclusive. The second structure states that the ultimate goal of the system is resilience, and that sustainability is a process that helps this goal. The third-way analyses resilience as a necessary precondition for the fulfillment of sustainability, considering that business activities are sustainable as long as their core activities are resilient to disruptive events in this sense, if the system lacks resilience, it can only possess a fragile sustainability according to Ahern (2013).

The relationship between lean practices and sustainability performance takes into consideration three measurable dimensions: environmental, social, and economic He et al. (2019). Lean manufacturing is inherently compatible with environmental practices, and a catalyst towards further environmental improvement Belhadi et al. (2018). Both Lean management and environmental management share a common objective, which is the elimination and prevention of all types of waste at the point of origin Martinez-Jurado and Moyano-Fuentes (2014).

Lean practices has been associated positively with social performance, since factors such as motivation, communication, team-work, and problem solving for continuous improvement are human aspects at the heart of the lean paradigm (Hines et al. 2004). Howerver, according to Anderson-Connolly et al. (2002), lean practices is not inherently antithetical to employees wellbeing, which ultimately depends on intensity. Furthermore, it has been suggested that the positive effect of lean systems on social performance depends on how potentially harmful lean manufacturing techniques are when combined with human factors to counteract possible negative effects as stated by Camuffo, DeStefano, and Paolino (2017). Thus, while the true spirit of lean seems to generally support employees wellbeing, the author require the relationship between Lean and social performance to be further investigated.

To Sum up, indeed these three paradigms have their respective objectives and practices, Lean focuses on the waste elimination, Resilient practices work towards how

SMEs respond to the unexpected event such as COVID-19 pandemic and Sustainability concept urges for embracing the triple bottom line concepts, which has all three dimensions, that is, economic, social and environmental. However, they serve for a common purpose, that is, to make SME's business more competitive, efficient and effective. Organizations and SMEs are encouraged to adopt Lean, resilient and sustainable approach initiatives, as a way of managing business to ensure competitive advantage. Table 8.2 shows the impact of this adoption.

Table 8.2 Impact of 'Resilient-Lean-Sustainable' approach on SMEs

Approach	Main practices	Impact on SMEs		References
		Positive	Negative	
Lean	Value stream mapping, JIT, Kanban, respect for employees, automated mistake proofing, 5S SMED and cellular manufacturing	• Competitive intensity of industry (cost leadership and differentiation) • Cost savings; improve plant utilization, service responsiveness and profit margin • First-pass correct output, reduced manufacturing lead time, and increased productivity (4)	• Plant layout changes • redefinition of work standards • Changes in the logic of production planning and control	• Düeset et al. (2013) • Hallgren and Olhager (2009) • Zhou (2012) • Ghosh (2013) • Carvalho et al. (2011)
Resilience	• Risk management culture • Lifeline maintenance • BDA-driven and real-time information system	• Low customer dependence and supplier dependence • Low supplier concentration and multiple/ global sourcing • Ability to absorb disruption and business continuity	• High inventory levels • High production costs	• Christopher and Peck (2004) • Wagner and Bode (2006) • Van der Vorst and Beulens (2002) • Kamble (2020) • Graveline and Grémont (2017)

(continued)

Table 8.2 (continued)

Approach	Main practices	Impact on SMEs		References
		Positive	Negative	
Sustainable manufacturing	• Long-term relationships, enhanced communication, logistics and technological integration, development operation standardization and selective monitoring, stakeholder management and life cycle assessment	• Regulatory compliance, market value, production cost • Environmental championing, monitoring (buyers), public and regulatory pressure • Financial benefit, company image, environmental conservation, compliance with regulations, stakeholders, green innovation, supply chain requirements, customer awareness, employee demands, internal motivations, market trend, competitor	• Complex processes • High costs	• Schaltegger and Burritt (2014) • Baines et al. (2012) • Lee and Klassen (2008) • Govindan et al. (2014) • Singh et al. (2016)

8.2.3 Challenges of Adoption Sustainable, Lean and Resilient Strategies in SMEs

Lean has been designed and confirmed by large companies, which have been able to achieve a significant increase in production capacity and performance (Belhadi et al. 2018). Furthermore, many researchers have promoted the use of lean in small and medium enterprises as well (Belhadi et al. 2018; Panizzolo et al. 2012) and outlined its applicability in such companies and the possible competitive advantages achieved through lean production in SMEs. However, a number of authors have argued that the types of organisations who have firstly embarked on the Lean journey and secondly found success in this have been predominately larger organisations (Shah and Ward 2003; Bhamu and Sangwan 2014). The adoption of lean in Small and Medium Enterprises (SMEs) pose even more challenges. As indicated by Achanga et al., SMEs require that the implementation costs and the subsequent benefits of lean manufacturing adoption, be projected upfront before they are able to commit. Further to this, SMEs compared to large organizations have limited resources, and in many cases the leadership lacks the long term commitment required.

Kumar and Kumar (2014) focused on the barriers in the implementation of lean manufacturing, and grouped them into seven categories: management, resource, knowledge, conflicts, employee, financial and past experience. Kumar and Kumar (2014) assumed that the lack of necessary resources (labour, capital, communication etc.) prohibit the implementation of lean manufacturing. Furthermore, lots of companies that attempt to introduce lean practices (and eventually thinking) rely on consultants, thus resources for consultancy are key as well. In addition, AlManei (2017) found that the quality of the consultant is also critical, and in many cases superficial knowledge of the subject and lack of implementation practices results into confusion about Lean Manufacturing and can become an obstacle in Lean Manufacturing implementation. It is thus evident that knowledge of the subject is of paramount importance as well. Absence of knowledge on lean philosophy and the various tools can be a great barrier in the implementation.

The lack of sustainability efforts in SMEs is attributed due to characteristics of SMEs. SMEs often lack the awareness, expertise, skills, finance, and human resources to build the required changes for sustainability within the organization [10, 14]. Hillary (2004) identified barriers and drivers for the environmental management system for SMEs. These barriers are lack of knowledge, training, implementation cost, transient cost and so on. The drivers for sustainability in SMEs, as identified by Hillary (2004), are customers, government, local community, employees, insurers, banks and larger companies. This study concluded that despite these barriers, SMEs do achieve benefits from Environmental Management System (EMS). Lepoutre and Heene (2006) reported that firm size and characteristics of SMEs are also recognized as barriers for sustainable practices. However, the effect of these barriers can be nullified by critical analysis and strategy to overcome the constraining barriers. The adoption of sustainable management led managers to be confronted with tensions and conflicts in the face of the different dimensions of sustainable development (Hahn et al. 2015). In particular, the structural lack of resources would often lead SMEs to see sustainable development issues as a constraint, rather than as a business opportunity. In fact, small businesses face certain obstacles that risk slowing them down in adopting sustainable practices, including, of course, the lack of resources, but also the lack of time, incentives and support for sustainable initiatives (Bonneveux, Calmé, and Soparnot 2012).

COVID pandemic represents an external shock of unprecedented magnitude, affecting SMEs on the supply and demand sides alike (Juergensen et al. 2020), because of their size and ownership structure, they also generally struggle with profitability and liquidity, thus becoming particularly vulnerable to external shocks (European Commission 2019). SMEs are generally less resilient compared to larger firms (Martin et al. 2019), meaning according (Juergensen et al. 2020) to that they take longer to return (if at all) to normal operations following a crisis. Across SMEs there is a lack of knowledge and awareness of the potential impacts of extreme weather, with many SME managers and decision-makers not realising how vulnerable their businesses are. Resilience can help the entrepreneur to withstand internal and external shocks and may be one of the drivers of entrepreneurial performance or success (Bullough et al. 2017). Found that resilience helps entrepreneurs to manage business failure and helps re-entry into entrepreneurship. In addition, resilience is positively linked to innovation, which is one of the attributes of successful entrepreneurship. High growth firms tend to be a rational consequence of innovation. Growth is a proxy for entrepreneurial success. A resilient business constantly changes its portfolio of innovation to adapt to changing environmental conditions (Markman et al. 2017). Examined the relationship between resilience and innovation of SMEs. The results indicate that there is a significant positive relationship between innovation as measured by idea management capability, strategic management capability, and knowledge management and resilience. Reinmoeller et al. (2008) also find a positive relationship between resilience and the success of the businesses. Ofunoye (2015) however found an insignificant relationship between resilience and business success as measured by employee numbers, profitability growth, and sales growth.

8.3 Case Study

This research aims to adopt Sustainable, lean and resilient strategies to hence SMEs during COVID pandemic. To fulfill this aim, The case study is considered to be the most relevant approach to theory development through the observation of constructs that develop in the real fields Eisenhardt (1991), especially for new topics Edmondson (2007). This is the perfect choice for this case.

Moreover, case studies are tailor-made for exploring new processes or behaviors or ones that are little understood (Hartley 1994). Hence, the approach is particularly useful for responding to how and why questions about a contemporary set of events (Leonard-Barton 1990).

Hartley (2004) states that case study research "consists of a detailed investigation, often with data collected over a period of time, of phenomena, within their context," with the aim being "to provide an analysis of the context and processes which illuminate the theoretical issues being studied". The purpose of using the case study is to get in-depth details as much as possible about an event, person or process Yin (2003).

Case studies can involve single or multiple cases. The problem of single cases is limitations in generalizability and several information-processing biases (Eisenhardt 1989). One way to respond to these biases is by applying a multi-case approach (Leonard and Barton 1990). Multiple cases augment external validity and help guard against observer biases. Moreover, multi-case sampling adds confidence to findings. By looking at a range of similar and contrasting cases, we can understand a single-case finding, grounding it by specifying how and where and, if possible, why it behaves as it does (Miles and Huberman 1994). Yin (1994) pointed out that generalization of results, from either single or multiple designs, is made to theory and not to populations. Multiple cases strengthen the results by replicating the pattern-matching, thus increasing confidence in the robustness of the theory.

Data Collection in Case study is one of the most significant activities in the process because the richness and depth of what will be eventually known is contingent on the craft and effectiveness of the data collection method in uncovering relevant details about the situation (Baboucarr and Soaib 2014). The choice of data collection procedures should be guided by the research question and the choice of design. The case study approach typically combines data collection methods such as archives, interviews, questionnaires, and observations (Yin 1989).

Interviews are one of the most important sources of case study information. There are several forms of interviews that are possible: Open-ended, Focused, and Structured or survey. In an open-ended interview, key respondents are asked to comment about certain events. They may propose solutions or provide insight into events. They may also corroborate evidence obtained from other sources. The researcher must avoid becoming dependent on a single information, and seek the same data from other sources to verify its authenticity Winston (1997).

8.3.1 Within Case Analysis

Case studies were carried out in SMEs companies in Morocco in order to analyze the environmental management level of the companies and their degree of LRS implementation during COVID pandemic. Case studies explore and study real-life phenomena through the detailed contextual analysis of a limited number of events or conditions and their relationships (Zainal 2007). A total of 38 companies were contacted by email or telephone, the selection criteria to choose the companies for the case study were: (i) relevance to understand of LRS practices; (ii) the companies should be small and medium size companies; (iii) willingness to take part in the study; and (iv) geographical proximity to ensure the right conditions for the research team. But only 4 were willing to participate in the case studies (Table 8.3).

The analysis of these companies was carried out in two different steps. First of all, an online survey was sent in order to assess the management level of each company. Once the answers were evaluated, face-to-face semi-structured interviews were conducted at the companies, with meetings and personal interviews which lasted

Table 8.3 Summary of the case companies

Company	Sector of activity	Turnover/balance	Number of employees	Age
Company A	Textile and leather	20 millions de Dhs	40	15
Company B	Agro processing	11 millions de Dhs	28	8
Company C	Construction	7 millions de Dhs	17	11
Company D	E-commerce	4 millions de Dhs	10	5

about two hours. This provided in-depth insights into the views of the corporate representatives and allowed us to gain rich contextual information regarding the implementation of the circular economy among the companies. Questions were open-ended in order to allow a deeper exploration of issues and to allow respondent's own experiences to emerge freely (Aira et al. 2003; Ingram 2008).

8.3.1.1 Case of Company A

Company A is a female business that was founded in 2005. It provides a variety of handmade high quality leather shoes, handbags and accessories with hand painted details.

- **Implementing LRS during COVID-19 in company A**

For Company A, the crisis led to the loss of loyal customers and the uncertainty of business. Company A has taken a series of practices. For example, Company A kept scanning the environment and detecting signals of potential crisis. Accordingly, Company A's managers proactively established and effectively executed relevant crisis management plans. Specifically, Company A diversified its market segments and purchased insurance. Company A switched its target market and reduced operational costs by laying off employees, shortening working time and consolidating production processes. During the last months, Company A adjusted their strategic plans and operational arrangements based on the lessons of the crisis events to improve their business effectiveness.

The company did not follow an established framework for lean manufacturing implementation. The general manager instructed the operations manager (also the team leader) to start implementing lean manufacturing practices. At the outset, a general training program on lean manufacturing principles was organized for the middle management and the operators.

After lean manufacturing implementation resulted in a few changes on the work floor. A few visual management techniques were initiated. There were initiatives taken by the management to collect data from the work floor, such as overall equipment efficiency, total productive maintenance, and rework. Some of the important benefits of lean practice implementation at this SME were reduced machine downtime, improved quality, and work environment. There was no initiative taken toward pull production. Similarly, the traditional production method and layout possess several bottlenecks for smooth flow, which requires massive investment and space for any change in layout. As a result, Company A did not only survive from the crisis, but also rejuvenated its business by converting this crisis event into a new development opportunity, while most of other competitors are still struggling.

Sustainable activities are driven in the company. The sustainable approach implies, integrating various sustainability aspects into the realm of the company's management. Over the past three years of the implementation, the company has further enhanced the management system and improved upon the environmental and social performance. The company's responsibility driven approaches fully encompasses the company's supply chain. For example the code of conduct signed by suppliers ensures compliance pertaining to hygiene, and environmental, social factors including human rights.

8.3.1.2 Case of Company B

Company B is a family business established in 2013 based on their tradition and inherited chocolate recipes from their ancestor. Today Company D produces more than 150 different kinds of confectionary (sweets) products. They also offer customized products for individual/custom orders.

- **Implementing LRS during COVID-19 in company B**

  During COVID pandemic, Company B adopted a set of resilient practices, including expending its business, diversifying its customers and renewing its infrastructures. Company B switched from its current target market and invested heavily on the technological innovation. Company B starts lean implementation in 2018, they put a lot of improvements into action during the lean manufacturing implementation. They made an assessment of downtime, rework, and delivery. After one year, they had been able to reduce downtime by 25% and rework by 5%, and they improved delivery performance. During the pandemic, the commitment of top management was exemplary in company B. It was assigned to implementing lean manufacturing and granted power to take quick decisions on the shop-floor. The operation manager actively participated in the lean manufacturing implementation and successfully guided the team. The motivation of employees and management for lean manufacturing implementation has been high. Even with crisis consequences, they still had their weekly meetings led by the team leader. Company B showed great teamwork and team feeling. The management put a lot of effort into team building. The employees celebrate company events and special days with their colleagues. As a result, Company B successfully stopped the sales decrease and maintained their major customers. The company's responsibility driven approaches fully encompasses the company's supply chain. For example the code of conduct signed by suppliers ensures compliance pertaining to hygiene, and environmental, social factors including human rights.

  Sustainability is integrated with the core fundamental of the company. Environmental and social sustainability practices were ingrained in the whole system of the company. The company also believes that strong governance is central to creating value on a sustainable basis. The company obtained both ISO 9001 quality certification and ISO 14001 environment certification. The company's process safety management system approach has enabled to adopt best in class safety practices and management system, manufacturing division have been complied with ISO 45001certification. The company considers all the stakeholders as a partner in success and remains committed to maximizing the stakeholder value. For example, the customers get high quality products at competitive prices and employees enjoy a high satisfaction level. These actions contributed to employee motivation and helped to increase the conditions of work and maintained their major customers.

8.3.1.3 Case of Company C

Company C is a family business that was founded in 2010. At the beginning the enterprise was focused in building log cabins, garages, pergolas and making wooden panelling, floor, slats, etc. Later, in 2016, the enterprise shifted their value proposition to the production of wood biomass.

- **Implementing LRS during COVID-19 in company C**

  Company C was the major supplier's suspension. In particular, the only supplier of their key raw material suddenly increased the price and then stopped supplying because of the crisis. This unexpected crisis forced Company C to stop current production line and switch to a much less profitable production line. To deal with this crisis, Company C did not detect any signal prior to the crisis events and had no contingency plan. It just reactively responded to this crisis event by switching the profitable target market. Company C was in lack of a systematic and effective resilience strategy. Company C started implementation at the same time of COVID crisis. The general owner is very motivated to implement lean principles. However, the operations managers were resistant to change. This imbalance in leadership and support for improvement hugely affected lean implementation. The general feeling in the company was that the production process is optimal and not much can be done. The observations and interviews provided clear evidence of the management's firefighting behavior. COVID pandemic eventually affected heavily Company C.

  During COVID crisis, the company started a sustainability management framework that has been introduced and further institutionalized to aid the planning and sustainable led initiatives across the operations. The framework has four designated stages such as, planning, implementation, checking and corrective actions and review. The company has instituted many commendable environmental friendly, sustainable activities in the entire supply chain.

8.3.1.4 Case of Company D

Company D was founded in beginning of 2015's as internet service provider and has transformed over the years into high tech IT service and solutions provider in the field of High-Performance Computing (HPC). Today it offers services of supercomputer infrastructure to their clients and provides them system administration, optimization and parallelization of code, cloud computing services, web and mobile application development services and project management SaaS services.

- **Implementing LRS during COVID-19 in company D**

  The financial and reputational loss is huge resulting from this crisis. In the entire process of crisis management, Company D did not have any pre-warning mechanisms and contingency plans. To deal with this crisis, Company D established a recovery plan to catch up all the delayed works and proactively communicate with customers in order to sustain the business with them. Company D initiated lean manufacturing with a formal communication from the top management to all the employees. The company made good progress in implementing lean principles over the year 2020 and improved operational performance in many respects. In the initial phase it started slowly because of the global crisis environment. However, after they approved implementation, it was quickly put into practice. The company was able to improve the product quality, reduce weight variation of the packets, and reduce costs by reducing rework. Moreover, total productive maintenance was initiated to reduce machine downtime. As a result, Company D eventually overcame the crisis event.

The company has started sustainable approach during COVID pandemic, the company in its sustainability activities addressed social issues such as health and safety, ethics, education to employees, sanitation, sustainable supplier development, and product responsibility for customers. the company is committed towards ethical working practices and human rights and complies with international, national and regional norms to fair and ethical working practices. The company employees are made aware of the expectations to not to engage in corrupt practices in their business dealings. In all sales and marketing operations, local staffs are hired, and they were trained in functional and soft skills aspects. For customers, the periodic customer satisfaction surveys are conducted and the feedback on product performance is obtained and the customer grievances are addressed.

8.3.2 Cross-Case Comparison

In the following section, we will discuss the similarities and differences between both the cases with regard to their implementation of LRS paradigms.

All the companies suffered heavily from the COVID pandemic, although company C seems to suffer the most because of it major supplier's suspension. Tables 8.4 and 8.5 highlighted the managerial and tools aspect of LRS paradigm on case companies. All the companies have integrated Lean approach in their policy and corporate strategy. Company B was the first firm to make the change towards LM in 2018 and several operational benefits have already been observed. The other companies begun to approach lean principles more recently or during the crisis (company C and D), so they are characterized by a lower lean implementation level. All the companies except C succeeded, whereas company C struggle to implement lean manufacturing due to the imbalance in leadership between general owner and operation manager and the lack of understanding the lean approach. In all the companies the requirement to change came from the top and then was expanded towards the bottom: some companies started to engage employees of all levels from the beginning (case A, B, D) while others (case C) involved them gradually after training managers first, with an expansion of the training from the top towards the bottom of the organizational pyramid. The methodologies that constitute the backbone of LM and that are implemented more often are 5S, TPM, VSM, Kaizen and JAT.

Table 8.4 Techniques and Tools of Lean Resilient Sustainable paradigms employed by the research units

	Company A	Company B	Company C	Company D
Lean	• 5S • JAT • KAIZEN • TPM • VSM	• 5S • JAT • TPM • VSM	• 5S • TPM	• 5S • JAT • TPM • VSM
Resilient	• Executing relevant crisis management plans • Switched target market • Laying off employees • Shortening working time • Consolidating production processes	• Expending business • Diversifying customers • Renewing infrastructures • Switched current target market • Invested heavily on the technological innovation	• Stop current production line • Switch to a much less profitable production line • Switching the profitable target market	• Establishing a recovery plan • Communicating with customers
Sustainable	• Integrating various sustainability aspects into company's management • Code of conduct with suppliers (hygiene, and environmental, social factors)	• ISO 14001 environment certification • Safety management system approach • Maximizing the stakeholder value	• Introducing framework for sustainable activities • Environmental friendly, sustainable activities in the entire supply chain	• Ethical working practices • Satisfaction of the client • Trained employees

Both company A and B did not detect any signal prior to the crisis events and had no contingency plan to face the crisis, during the pandemic they adopted some resilience practices including switching the target market and establishing a recovery plan. Whereas, the company A and B have detected signals of potential crisis and have prepared a multiple resilience practices to deal with the pandemic such as expending business, diversifying customers and renewing infrastructures.

All the companies involved in this study have clearly designed their sustainability goals and they have already implemented social or environmental actions. All the companies have clear idea about what their impact is on the surrounding environment and on the social community they operate in. Table 8.6 underlined the impact of Lean-Resilient-Sustainable implementation on companies (A, B, C and D).

Table 8.5 Managerial aspects of LRS paradigms implementation

	Company A	Company B	Company C	Company D
Lean	• The company did not follow an established framework for lean manufacturing implementation • The general manager instructed the operations manager to start implementing lean manufacturing practices • A general training program on lean manufacturing principles was organized for the middle management and the operators	• Top management assigned to implementing lean manufacturing • The operation manager actively participated in the lean manufacturing implementation and successfully guided the team	• The general owner is very motivated to implement lean principles • The operations managers were resistant to change • The management's firefighting behavior affected the implementation	• Initiated lean manufacturing with a formal communication from the top management to all the employees • The company made good progress in implementing lean principles
Resilient	• Scanning the environment and detecting signals of potential crisis • Proactively established and effectively executed relevant crisis management plans	• Any crisis plan before the pandemic • Company B adopted a set of resilient practices during the crisis	• Any signal detected prior to the crisis events and no contingency plan • Lack of a systematic and effective resilience strategy during the crisis	• Any pre-warning mechanisms and contingency plans
Sustainable	• Sustainability integrated upon the environmental and social performance	• Environmental and social sustainability practices were ingrained in the whole system of the company	• The company started a sustainability management framework regrouping both social and environmental issues	• Sustainability activities addressed social issues to employees and product responsibility for customers

Table 8.6 Impact of LAR paradigms implementation on companies (A, B, C and D)

	Company A	Company B	Company C	Company D
Lean	• Reduced machine downtime • Improved quality, and work environment	• Reduce downtime by 25% and rework by 5% • Improved delivery performance	• Financial savings • Less process Waste	• Improve the product quality reduce weight variation of the packets • Reduce costs by reducing rework. reduce machine downtime
Resilient	• Improve business effectiveness	• Improve business effectiveness	• Global sourcing	• Sustain the business with costumers
Sustainable	• Enhanced the management system	• Employee motivation Increase the conditions of work • Maintained the major customers	• Company image • Internal motivation	• Employees are made aware of practices • Client satisfaction

8.4 Framework for Adopting Lean-Resilient-Sustainable Paradigm in SMEs During COVID-19 Pandemic

After examining the limitations of the existing studies, evaluating the strengths associated with the Lean, resilient and sustainable strategies, and investigating the drivers, enablers, outcomes, Critical success factors, and barriers of LRS, it can be concluded that a combined LRS approach is needed for optimal SME Business sustainability benefits to face post COVID pandemic. Figure 8.1 presents a LRS framework comprising an integrated LRS approach at the core in combination with theoretical process constructs including the key drivers, enablers, critical success factors, barriers, and outcomes.

The integrated framework facilitates the adoption of lean, resilient, and sustainable strategies in overcoming the limitations of LRS in achieving SME benefits outcomes. The proposed framework is an arrangement of the LRS components that strengthen the connection and relationship between these three strategies. This framework has been designed after an extensive literature review and evaluation of the case study. It consists of three stages which lead one to the other. First, the drivers that motivate SME to implement the LRS paradigms. In the next stage, the tools help an SME to execute and achieving LRS implementation. The arrows in the central part of the model depict the support from lean, resilient, and sustainable in overcoming the shortcomings of these strategies in different combinations to emerge as an integrated strategy. Finally, in the last stage, the outcomes are achieved by the actions taken utilizing the strategy. In addition to this framework, the Critical success factors play

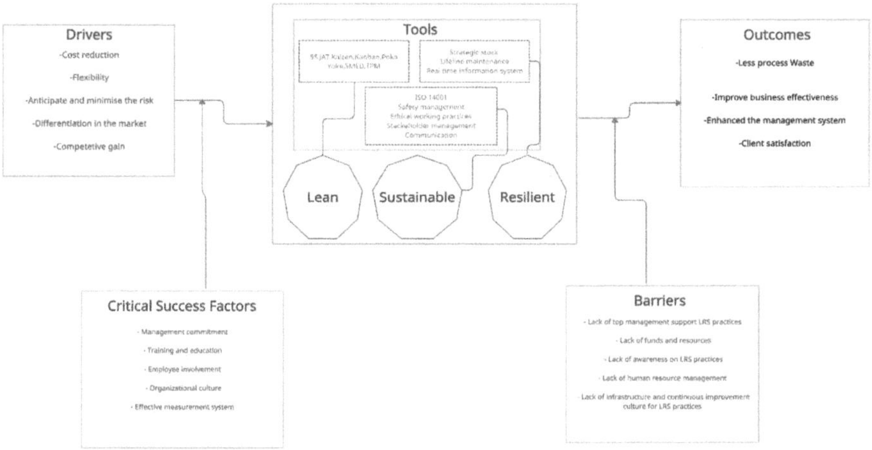

Fig. 8.1 LRS approach Framework towards SME sustainability

an intermediate role to ensure the effective implementation of the LRS. The barriers perform as impeding factors to impact the implementation and must be overcome by developing policies and guidelines. The success of this combined approach depends on an assessment of the drivers, tools, critical success factors, barriers, and outcomes. These components work together to facilitate the implementation within SMEs.

8.5 Conclusion

The Lean, resilient and sustainability are leading techniques of manufacturing. There is no doubt that each paradigm has its immense application in their own regime and its worth understanding. In order to understand these techniques, the focus of this study is the utilization and union of available previous related literature related to all these techniques (lean, resilient and sustainability) on one podium for understanding and generation of their interconnection with each other. The literature study revealed that the lean has a typical focus on the detection and reduction in wastes (non-value added activates). The aptitude and ability of any system to pact with risks and disruptions without affecting quality and quantity of output and quantity of products considered as resiliency. Sustainability is helpful but requires optimized technology and improved policies to minimize resource usage and toxic wastes and improved product lifecycle. Hence, this research work is an addition to the literature study that is supportive to bring out clarity that the combination of all these processes is very helpful for the sustainable production and the application of LRS idea in combination.

This study has helped in investigating the theoretical constructs—drivers, enablers, barriers, and outcomes for an integrated LRS approach to enhance SMEs business in post COVID phase, through a systematic literature review and the multiple case studies. The exploration of these constructs provides deep insights into the impact of a LRS strategy on utilizing resources, adding value, managing wastes, and the resulting sustainability. This study has emphasized how a combined LRS approach can be useful.

The main limitation of the present research work is that the LRS framework has not tested pragmatically. This limitation provides the impetus for future research to implement LRS in different sectors. Future research can also focus on the role of LRS for sustainability enhancement through industry 4.0 and modeling and investigation of LRS barriers.

References

Ahern J (2013) Urban landscape sustainability and resilience: The promise and challenges of integrating ecology with urban planning and design. Landsc Ecol 28:1203–1212

Anand G, Kodali R (2008) A conceptual framework for lean supplychain and its implementation. Int J Value Chain Manage 2:313–357

Andersson P, Aronsson H, Storhagen NG (1989) Measuring logistics performance. Eng Costs Prod Econ 17:253–262

Asgary A, Ozdemir A (2020) Small and medium enterprises and global risks: evidence from manufacturing SMEs in Turkey. Int J Disaster Risk Sci 11:59–73

Belhadi A, Touriki FE, El Fezazi S (2018) Benefits of adopting lean production on green performance of SMEs: A case study. Prod Plan Control 29(11):873–894

Carvalho H, Barroso AP, Machado VH, Azevedo A, Cruz-Machado V (2012) Supply chain redesign for resilience using simulation. Comput Ind Eng 62(1):329–341

Fong, M (2020) Nonpharmaceutical measures for pandemic Influenza in Nonhealthcare settings—social distancing measures 26(5):976–984

Giannacouroua M, Kantaraki M, Christopoulouc V (2015) The perception of crisis by Greek SMEs and its impact on managerial practices. Procedia Soc Behav Sci 175:546–551

Hamid RA (2011) Factors influencing the success of lean services implementation: conceptual framework. 2nd ICBER. Langkawi Kedah, Malaysia

Haut-Commissariat au Plan (2020). IMPACT SOCIAL & ECONOMIQUE DE LA CRISE DU COVID-19 AU MAROC, Juillet 2020

He Q, Gallear D, Ghobadian A, Ramanathan R (2019) Managing knowledge in supply chains: A catalyst to triple bottom line sustainability. Prod Plan Control 30 (5–6):448–463

Hines P, Holweg M, Rich N (2004). Learning to evolve: a review of contemporary lean thinking. Int J Oper Prod Manag 24(10):994–1011.

International Monetary Fund (2020, April) World economic outlook, April 2020: The Great Lockdown. IMF.

International Trade Centre (2020) SME competitiveness outlook: COVID-19: the great lockdown and its impact on small businesses. ITC, Geneva, p 33

Jenkins H (2009) A "business opportunity" model of corporate social responsibility for small-and medium-sized enterprises. Business Ethics: A European Review, 18(1):21–36

Juergensen J, Guimón J, Narula R (2020) European SMEs amidst the COVID19 crisis: assessing. J Ind Bus Econ 47:499–510

Kainuma Y, Tawara N (2006) A multiple attribute utility theoryapproach to lean and green supply chain management. Int J Prod Econ 101(1):99–108

Kechiche A, Soparnot R (2012) CSR within SMEs: literature review. Int Bus Res 5(7):97–104

Kumar R, Kumar V (2014) Barriers in implementation of lean manufacturing system in Indian industry: A survey. Int J Latest Trends Eng Technol 4(2):243–251

Lengnick-Hall C, Beck T (2003) Beyond bouncing back: the concept of organizational resilience. Paper presented at the Academy of Management, Seattle, WA, 1–6 August.

Marchese D, Reynolds E, Bates ME, Morgan H, Clark SS, Linkov I (2018) Resilience and sustainability: similarities and differences in environmental management applications. Sci Total Environ 613–614:1275–1283

Martinez-Jurado PJ, Moyano-Fuentes, J (2014) Lean management, supply chain management and sustainability: a literature review. J Clean Prod 85:134–150

Munoz Morezo R (2020, June 15) How to support MSMEs so they overcome the COVID-19 shock. Folha de São Paulo (in Portuguese)

OECD (2009) The impact of the global crisis on SME and entrepreneurship financing and policy responses.

Ohno T (1998) The Toyota production system. Productivity Press, Portland, OR

Qiu Y (2020). Impacts of social and economic factors on the transmission of coronavirus disease 2019 (COVID-19) in China 494

Reichhart A, Holweg M (2007) Lean distribution: concepts, contributions, conflicts. Int J Prod Res 45(16):3699–3722

Rosič H, Bauer G, Jammernegg W (2009) A framework for economic and environmental sustainability and resilience of supply chains. In: Rapid modelling for increasing competitiveness, pp 91–104

Seuring S, Müller M (2008) From a literature review to a conceptual framework for sustainable supply chain management. J Clean Prod 16:1699–1710

Seuring S (2013) A review of modeling approaches for sustainable supply chain management. Decis Support Syst 54:1513–1520

Shefi M, Liu J, Ren W (2020) Impact of COVID-19 pandemic on micro, small and medium-sized enterprises operating in Pakistan. Research in Globalization Elsevier

Sterling S (2011) Encouraging resilience within SMEs: the cabinet office's proposed approach. J Bus Contin Emer Plan 5(2):128–139

Van der Vegt GS, Essens P, Wahlström M, George G (2015) Managing risk and resilience. Acad Manag J 58(4):971–980

Vonderembse MA, Uppal M, Huang SH, Dismukes JP (2006) Designing supply chains: towards theory development. Int J Prod Econ 100:223–238

Wildavsky A (1988) Searching for safety. Transaction Books, New Brunswick, CT

World Trade Organization (WTO) (2016) World Trade Report 2016: levelling the trading field for SMEs. WTO, Geneva, 2

Zeidy IA (2020) Economic impact of COVID-19 on MSMEs in Africa and policy options for mitigations. COMESA 9, 7

Chapter 9
Circular Economy in SMEs: The Role of Lean, Lean Six Sigma and Smart Manufacturing

Objectives

The aim of this chapter is to:

1. Highlight the relationship between circular economy, lean and lean six sigma
2. Highlight the relationship between circular economy and smart manufacturing
3. Highlight he integration between Circular Economy, Lean, Lean six sigma and smart manufacturing

9.1 Introduction

Recent studies have highlighted the crucial role of Small and Medium-Sized Enterprises (SMEs) as a driving force of the world economy (Belhadi et al. 2018). SMEs make up for more than 90% of the world's businesses while employing more than 60% of the world's population (Boddy et al. 2000). On the other hand, SMEs are responsible for a high consumption of resources, air and water pollution and waste generation (de Sousa Jabbour et al. 2020). Therefore, SMEs need to shift their traditional ways of doing business and adopt sustainable modes of manufacturing while remaining competitive. In this vein, circular economy (CE) emerges as a new philosophy that restores the damages caused in the resources acquisition phase, to generate minimum waste, to replace 'End of life' concept with restoration, to use renewable energy and to optimize the production process and product life cycle (MacArthur 2013).

Likewise, lean manufacturing (LM) and lean six sigma (LSS) came to light as management philosophies that would answer the efficiency/waste paradigms. Across the literature, there is a mutual agreement that LM and LSS have a promising potential in leveraging the CE principles in SMEs. Lean and LSS were found to contribute to environmental well-being, waste minimization and energy and resource savings as well as productivity, efficiency and cost savings (Taddeo et al. 2019). As a result,

F. E. Touriki et al., *Sustainable Excellence in Small and Medium Sized Enterprises*, Industrial Ecology, https://doi.org/10.1007/978-981-19-0371-7_9

SMEs should naturally adopt both LM /LSS and CE, as both strategies complement and support each other.

Smart Manufacturing (SM) or Industry 4.0 (I4.0) is another paradigm that has been recently added to the picture. Past literature argues that organizations, undergoing a transition from LE to CE, must implement technological development using the technologies of SM (Romero et al. 2021). SM can act as an enabler of both CE and LM/LSS. Both principles can be integrated into the same I4.0 environment in order to improve environmental and economic performance of SMEs (Ciliberto et al. 2021).

Accordingly, the aim of this chapter is to understand how LM, LSS and SM can enable the strategies of CE in SMEs. The answer to this question can help SMEs to better apprehend the benefits of CE that can be supported through lean principles in the SM environment.

9.2 Key Concepts

9.2.1 Circular Economy

The Ellen MacArthur Foundation, first defined CE as "an industrial economy that is restorative or regenerative by intention and design". Another definition was proposed by (Geng and Doberstein 2008) describing CE as "the "realization of a closed loop material flow in the whole economic system". A more holistic definition is the one suggested by (Geissdoerfer et al. 2017) defining CE as "a regenerative system in which resource input and waste, emission, and energy leakage are minimized by slowing, closing, and narrowing material and energy loops". Linguistically, CE is the antonym of linear economy (LE).

LE on the other hand is defined as "converting natural resources into waste, via production. Such production of waste leads to the deterioration of the environment [...] by the removal of natural capital from the environment [...] and by the reduction of the value of natural capital caused by pollution from waste" (Murray et al. 2017). Therefore, CE comes as a replacement to LE, envisaging to restore the damages caused in the resources acquisition phase, to generate minimum waste, to replace 'End of life' concept with restoration, to use renewable energy and to optimize the production process and product life cycle (MacArthur 2013), which can be achieved through long-lasting design, maintenance, repair, reuse, remanufacturing, refurbishing, and recycling (Geissdoerfer et al. 2017).

Suárez-Eiroa et al. (2019) discussed the theoretical strategies related to the concept of CE and proposed seven operational principles with practical strategies for the implementation of these CE as shown in Table 9.1.

Several authors, academics, practitioners have tried to define CE without reaching a consensus. After analyzing 114 definitions, the authors (Kirchherr et al. 2017) proposed a new one, defining CE as an economic system that replaces the concept of "end-of-life" with reduction, alternative reuse, recycling and recovery of materials

Table 9.1 Practical strategies of Circular Economy (Suárez-Eiroa et al. 2019)

Strategies
Principle 1: Adjusting inputs to the system to regeneration rates • Substituting non-renewable by renewable inputs (e.g., bio-based materials, renewable energy) • Substituting renewable materials with low regeneration rates for other with faster regeneration rates • Adjusting taxes and subsidies of technology, products and materials based on their resource regeneration rates • Saving energy and materials (i.e., improving energy efficiency, resource productivity, virtualizing products, etc.) • Fostering renewable mobility (i.e., walking, bicycle, renewable fuels, etc.)
Principle 2: Adjusting outputs from the system to absorption rates • Substituting materials and processes which produce technical outputs by those which produce biological outputs • Substituting processes for those with lower waste generation rates (i.e. more eco-efficiency processes) • Adjusting taxes and subsidies of technology, products and materials based on their waste generation rates
Principle 3: Closing the system Separating biological and technical wastes properly • Re-manufacting products and components • Promoting and improving downcycling, recycling and upcycling of wastes (i.e. logistics, take-back systems, technology, etc.) • Promoting energy recovery by converting waste into heat, electricity or fuel • Promoting Extended Producer Responsibility
Principle 4: Maintaining resource value within the system • Interconnecting stages (i.e. redistributing second-hand goods) • Promoting industrial symbiosis (i.e. establishing standards, cascading, by-products, etc.) • Increasing durability (i.e. practical guides for reparability, preventive and corrective maintenance, repurposing, etc.) • Reducing obsolescence (i.e. updating software)
Principle 5: Reducing the system's size • Informing consumers properly (i.e. eco-labelling, product labelling, product declarations, etc.) Expanding the Extended Consumer Responsibility Promoting functional service economy and sharing economy (i.e. collective mobility) • Promoting green procurement (i.e. local products, season products, etc.) • Adjusting selling doses to consumer doses
Principle 6: Designing for circular economy • Eco-design (i.e. optimizing packaging, improving durability, etc.) • Designing transparent, reproducible and scalable products to build the same products in other places based on local resources • Thinking about practical utilities and consumer preferences (customization/made to order) Designing new business models and strategies • Designing new methodologies to guarantee a continual improvement • Designing projects to promote sustainable development and circular economy
Principle 7: Educating for circular economy • Adjusting educational curricula to the current challenges • Promoting knowledge, skills, capabilities and values that ensure the proper performance of circular economy • Promoting habits and individual actions in favor of circular economy

in the production/distribution and consumption processes. This is an operational definition that facilitates the understanding of the practices to be undertaken when applying this EC. As for the Agency for the Environment and Energy Management (ADEME in (TEULON 2015)) defines CE as "an economic system of exchange and production that, at all stages of the product life cycle (goods and services), aims to increase the efficiency of resource use and decrease the impact on the environment". This definition encompasses both the internal perspective associating the CE to a production system and the external perspective connecting the company to its network during the whole life cycle of the product. We can deduce that it is an economy of means, i.e. fighting against all forms of waste, whatever the stage of the product's life, through responsible production and responsible consumption.

Many times, CE is used to achieve sustainability by eliminating the use of toxic materials, reusing, and eliminating waste through the specific implementation of a set of models, principles, and practices throughout the life cycle of products and systems (Rajput and Singh 2019; Korhonen et al. 2018). Some authors (Lüdeke-Freund et al. 2019) see CE not as an ultimate goal, but as an earlier part of an overall strategy to increase resource effectiveness and efficiency. A practical approach aimed at the sustainable use of finite natural resources by increasing the efficiency of resource use in order to contribute to a balance between Economy, Environment and Society (Aranda-Usón et al. 2020).

Certainly the "environment aspect is well present" as it is agreed that the circular economy is considered a general concept characterizing the integration of environmental wellbeing into economic activity (Pieroni et al. 2019), through the realization of an advantageous system of restoration or regeneration by consciously closing the loop of the linear life cycle of products and replacing the concept of end of life with restoration (Millar et al. 2019).

However, according to a study by (Kirchherr et al. 2017), the most important and most frequently mentioned objective by CE practitioners is economic prosperity thus considering CE as a means to stimulate growth. Followed by environmental quality. As for the third dimension of sustainable development, the results of the study confirm that social equity is rarely taken into account or even absent.

Another way of understanding the concept of EC is that of (ADEME n.d.) which proposes a scheme composed of three action areas and seven pillars associated with them. The fields of action are:

- **Supply and economic actors**: extraction/exploitation and sustainable purchases, ecodesign (products and processes), industrial and territorial ecology, economy of functionality.
- **Consumer demand and behavior**: responsible consumption (purchase, collaborative consumption, use), extension of the duration of use (re-use, repair, reuse);
- **Waste management:** recycling (material and organic).

9.2.2 Lean Manufacturing

LM evolved to be a successful manufacturing strategy all over the globe. Obviously, waste reduction is the core objective of LM. It stands for "anything other than the minimum amount of equipment, materials, parts, space, and time that are essential to add value to the product" (Russell and Taylor 1999). The sources for the non-value-added activity wastes are Transportation, Inventory, Motion Waiting, Overproduction, Over processing and Defects (Sundar et al. 2014). The benefits of lean manufacturing are evident. Both scholars and partitioners report the following advantages (Cherrafi et al. 2016; Belhadi et al. 2018, 2020; Sundar et al. 2014; Pavnaskar et al. 2003).

- Improved product quality.
- Reduction in cycle time, work in progress, inventories and costs.
- Reduction in defects and reworks.
- Improved on time deliveries, flexibility, productivity, capacity space and machinery utilization.
- Skills enhancement.

Over the years, many LM tools and techniques have been developed. Table 9.2 highlights key elements of LM.

9.2.3 Lean Six Sigma

LSS evolved to overcome the shortcomings of both LM and capitalize on their strengths. This integration aims to make use of every opportunity for improvement, where LM eliminate non-value-added activities and SS ensures process control and conformance (Arnheiter and Maleyeff 2005).

LSS achieves product quality, high customer service level, training and education of the workforce (Cherrafi et al. 2016; Albliwi et al. 2015). Furthermore, it creates an ongoing culture for continuous improvement, upgrades lean through more scientific approaches to quality and advances SS by adopting wider systems approach (Arnheiter and Maleyeff 2005). In the LSS methodology, SS complements lean by providing the necessary tools for handling manufacturing problems and non-conformances identified through lean application (Pepper and Spedding 2010).

Despite the evident benefits of LSS, it has been criticized across the literature. Harrison (2006) suggested that if the two concepts of LM and SS are implemented in isolation, they can neither reach their initial objectives neither the LSS objectives. Moreover, Bendell (2006) argued that LSS is "ill defined", therefore companies fail to provide effective training programs that responds to both LM and SS. Finally, the initiatives of LSS would be jeopardized if enterprises fail to reach an equilibrium

Table 9.2 LM key elements

LM elements	Reference
VSM "Each and every activity including Value-Added activity (VA) and Non-Value-Added activity (NVA) required to convert the raw material into finished product through the mapping of process and information flows essential to every product"	(Rother and Shook 2003)
Kanban/Pull The Pull system rely on customer requirement whereas push system rely on predetermined schedule	(Dengiz and Akbay 2000)
SMED systematic reduction of changeover time by converting possible internal setting time to external time	(Dillon and Shingo 1985)
Standardized Work "The safest and most effective method to carry out a job in the shortest repeatable time as a result the utilization of resources such as people, machines, and material is effective"	(Sundar et al. 2014)
Cellular manufacturing "The facility grouping in order to produce the product with minimum process time, waiting time, and transportation by smoothen the process flow"	(Sundar et al. 2014)
Continuous Improvement (Kaizen) "The management driven element which effort the cultural change in the workplace"	(Sundar et al. 2014)
Total Production Maintenance (TPM) "Using machines, equipment, employees and supporting processes to maintain and improve the integrity of production and the quality of systems"	(Hashim et al. 2012)

between LM and SS "moving away from a blinkered approach in any one direction, risking becoming too lean and therefore rigid in responses to the market and subsequently impacting on value creation" (Pepper and Spedding 2010).

9.2.4 Smart Manufacturing

The National Institute of Standards and Technology defines SM as "a fully integrated, collaborative manufacturing system that responds in real time to meet changing demands and conditions in the factory, in the supply network and in customer needs".

"SM defines a set of practices that respond to a new wave of networked data and information technology capability destined to shape future manufacturing operations" (Davis et al. 2012).

Accordingly, across the literature, there is generally no accepted definition of SM (Kusiak 2018). However, both scholars and practitioners agree on the features, technologies and objectives of SM. In the SM environment, machines are fully connected, monitored and controlled through wireless networks, smart sensors and advanced computational intelligence (Wang et al. 2018).

Moreover, in this environment, data is considered to be a necessary condition, therefore manufacturers are beginning to realize the importance of data for the success of SM (Tao et al. 2018). SM integrates a variety of technologies ranging from big data analytics (BDA), the internet of things (IoT), cyber physical systems (CPS), artificial intelligence (AI), robotics, RFID technology, additive manufacturing (AM) control simulation and predictive engineering (Kusiak 2018). These technologies are proven to improve product and service quality, productivity, flexibility, efficiency, reduce waste and costs and promote high product customization in a sustainable way. In addition, they allow for a better resource utilization and improved delivery times. Into the bargain, they facilitate the decision-making process thanks to the available data and the tools. Furthermore, the SM technologies can monitor and control machines, equipment and even product through a cycle of feedback while collecting a huge amount of data, which can help correcting and preventing any potential troubleshooting (Ivanov et al. 2019; Kusiak 2019, 2018; de Sousa Jabbour et al. 2018).

In a nutshell, SM will leverage the traditional manufacturing to an intelligent and efficient environment with more informed decision-making, leading to dramatic business transformation, enhanced customer satisfaction, demand driven supply chain service and improved workforce involvement (Davis et al. 2012).

9.3 Circular Economy, Lean and Lean Six Sigma for SMEs

Across the literature, there is a mutual agreement that LM and LSS have a promising potential in leveraging the CE principles in SMEs. Lean and LSS were found to contribute to environmental well-being, waste minimization and energy and resource savings (Taddeo et al. 2019). Other authors like Fatemi and Franchetti (2016) and de Oliveira Furukawa, Cunha, and Pedreira (2016) registered a reduction in environmental costs, packaging and plastic bags after the implementation of lean and LSS. Visual stream mapping (VSM) is one of the examples where LM enhances green operations as it can be used as an environmental assessment tool in order to understand and identify environmental waste and value in the production process (Ciliberto et al. 2021).

Furthermore, LSS initiatives can increase the awareness of customers, stakeholders and employees about waste and provide tools to understand causes and effects of environmental issues (Garza-Reyes 2015).

Arroyo and Gonzalez (2016) argued that the definition of waste in the context of lean should be reconfigured in a way that integrates environmental and social impacts. This reconfiguration can be the first step to adopt a CE strategy, where LM can support eco-design, remanufacturing, recyclability and less energy consumption (Ciliberto et al. 2021). According to Nadeem (2019), the combination between LSS and CE seems natural as both paradigms complement each other. LSS and CE have different focus, however, their two major common elements of 'waste elimination' and 'value creation' make them synergistic to benefit manufacturers.

VSM is again one of the examples of how CE focus on improving natural capital can be achieved through lean as it can map the resource and create flow within the closed loop system. On the other hand, thanks to the pull production as one of the principles of lean, resources optimization can be improved by only producing what is ordered by customers (Nadeem 2019).

Typically, LSS does not directly highlight opportunities to optimize resources, however it yields the way for identifying activities that mainly focus on eliminating waste; therefore, manufacturers especially SMEs can consider it as a tool for CE (Ciliberto et al. 2021).

Ghobakhloo and Fathi (2019) suggested that many SMEs are now relying on lean digitized strategies to easily identity these opportunities. In a nutshell, CE, lean and LSS are compatible, their integration seems only natural to create solutions that minimize resources consumption and enhance the value-added to the end user (Romero and Rossi 2017).

9.4 Circular Economy and Smart Manufacturing for SMEs

Authors, throughout the literature, agree that organizations, undergoing a transition from LE to CE, must implement technological development using the technologies of SM or industry 4.0 (I4.0) (Romero et al. 2021). According to Okorie et al. (2018) SM can act as an enabler of the CE; furthermore, companies cannot avoid considering the technologies of SM when moving towards CE.

SM offers a variety of innovative ways for the monitoring and exploitation of natural resources and life cycle management (Rosa et al. 2020). AM was found to upgrade current recycling processes, support the remanufacturing of products and components and enhance the reuse of materials (Clemon and Zohdi 2018; Paris and Mandil 2017; Kravchenko et al. 2020). The features of AM lead to reduced use of materials, advance renewable and sustainable production and encourage the recycling of small wastes (Despeisse et al. 2017).

Moreover, CPS is another technology that can enable an improved lifecycle management, in addition to the development of new services (Caggiano 2018). Into the bargain, IoT can also be used to support new waste management strategies and provide dynamic feedback control loops (Hatzivasilis et al. 2018; Esmaeilian et al. 2018). The use of IoT sensors allow the monitoring of performance, therefore manufacturers can proactively ensure improved services and products. As a result, product life cycles are extended and 3R (reduce, re-use, and recycle) strategies are upgraded (de Sousa Jabbour et al. 2018). Furthermore, simulation can assist manufacturers in modelling material flows (Schäfers and Walther 2017) in the remanufacturing of products (Kuik et al. 2016) and in the improvement of natural resources exploitation as elaborated in the study of Rönnlund et al. (2016).

Cloud manufacturing has the ability to leverage on demand manufacturing services, besides its potential in the enhancement of the share business model of CE. Additionally, cloud platforms can help collect data on consumers behavior which can be used to improve product and service quality and design for improved utilization or replacement of equipment (Rymaszewska et al. 2017).

According to Stock and Seliger (2016), the technologies of SM contribute to sustainable operations, informed decision making for CE, and new business models creation which is in line with the principles of CE; therefore, we can conclude that SM offers great potential for the transition towards CE.

9.5 The Integration of Circular Economy, Lean, Lean Six Sigma and Smart Manufacturing

Throughout the literature, there is a mutual agreement that the era of digital manufacturing holds great potential in promoting both LM/LSS and CE for SMEs.

The positive impact of SM on LM and LSS is evident. The barriers that face SMEs in the implementation of lean can be overcome trough SM (Kolberg and Zühlke 2015). In addition, SM will improve performance, flexibility, cost and quality which perfectly support the lean principles (Moeuf et al. 2018). Furthermore, the set of technologies and smart solutions provided by SM will enable SMEs to become lean without "striving-for-lean" effort (Sanders et al. 2016). Into the bargain, six sigma approaches can be greatly enhanced thanks to the power of advanced analytics of SM (Arcidiacono and Pieroni 2018). Therefore, managers can make well informed decisions for quality problems.

On the other hand, the adoption of lean under the umbrella of SM for CE will help improving efficiency, productivity, flexibility, transparency, and reduce complexity. "Both CE principles and lean strategies can be integrated into the same I4.0 environment to improve the competitiveness of organizations" (Ciliberto et al. 2021). Lean will improve the qualitative and quantitative measurement and analysis of consumption in term of resources and energy while paying attention to recovery and reuse of secondary raw materials (Ioppolo et al. 2014). CE and its sustainable strategies will improve the SME's competitiveness and help enhancing the customer and stakeholders' relationships, which will further foster the lean implementation.

In the SM environment, process flow is simplified and intelligent, which will reduce lead times and deliveries; as a result, waste is reduced and resources are optimized (Stahel 1997). According to Reche et al. (2020), the digitalization of manufacturing processes has a crucial role in the extension and sustainability of innovative management strategies that move beyond the "cradle to cradle" strategy.

In a nutshell, the new technologies of SM and lean principles offer a galaxy of opportunities for SMEs to enhance their sustainable production, which will make it easier for them to shift towards CE while remaining competitive.

9.6 Conclusion

In this chapter we shed some light on the paradigms of CE, LM, LSS and SM in the context of SMEs. It is clear that SM hold enormous potential in supporting the strategies of both LM/LSS and CE. Into the bargain, LM/LSS positively correlate with CE as both paradigms have the same objectives despite their different approaches.

The added value of this paper is a better understanding of how SM can promote CE strategies in a lean environment. The digital business transformation acting as the base that enables the development and implementation of CE both directly and indirectly through LM/LSS. In a nutshell this work can motivate managers and practitioners of SMEs to consider the adoption of CE, LM/LSS and SM through the apprehension of the potential benefits and opportunities.

References

ADEME Ad (n.d.) Consulté le Février 2021, sur https://www.ademe.fr/
Albliwi SA, Antony J, Abdul halim Lim S (2015) A Systematic Review of Lean Six Sigma for the Manufacturing Industry. Bus Process Manag J
Aranda-Usón A, Portillo-Tarragona P, Scarpellini S, Llena-Macarulla F (2020) The progressive adoption of a circular economy by businesses for cleaner production: An approach from a regional study in Spain. J Clean Prod 247
Arcidiacono G, Pieroni A (2018) The revolution lean six sigma 4.0. Int J Adv Sci, Eng Inf Technol 8(1): 141–49. doi:https://doi.org/10.18517/ijaseit.8.1.4593
Arnheiter ED, Maleyeff J (2005) The integration of lean management and six sigma. The TQM Magazine
Arroyo P, Gonzalez V (2016) Rethinking waste definition to account for environmental and social impacts. In Annual Conference of the International Group for Lean Construction. Vol. 24. Citeseer
Belhadi A, Kamble SS, Zkik K, Cherrafi A, Touriki FE (2020) The integrated effect of big data analytics, lean six sigma and green manufacturing on the environmental performance of manufacturing companies: the case of North Africa. J Clean Prod 252. https://doi.org/10.1016/j.jclepro.2019.119903
Belhadi A, Sha'ri YBM, Touriki FE, El Fezazi S (2018) Lean production in SMEs: literature review and reflection on future challenges. J Ind Prod Eng 35(6). Taylor & Francis: 368–382
Bendell T (2006) A review and comparison of six sigma and the lean organisations. Emerald Group Publishing Limited, The TQM Magazine
Boddy D, Macbeth D, Wagner B (2000) Implementing collaboration between organizations: an empirical study of supply chain partnering. J Manag Stud 37(7). Wiley Online Library: 1003–1018
Caggiano A (2018) Cloud-based manufacturing process monitoring for smart diagnosis services. Int J Comput Integr Manuf 31(7). Taylor & Francis: 612–623
Cherrafi A, Elfezazi S, Chiarini A, Mokhlis A, Benhida K (2016) The integration of lean manufacturing, six sigma and sustainability: a literature review and future research directions for developing a specific model. J Clean Prod 139. Elsevier: 828–846
Ciliberto C, Szopik-Depczyńska K, Tarczyńska-Łuniewska M, Ruggieri A, Ioppolo G (2021) Enabling the circular economy transition: a sustainable lean manufacturing recipe for industry 4.0. Business Strategy and the Environment. Wiley Online Library
Clemon LM, Zohdi TI (2018) On the tolerable limits of granulated recycled material additives to maintain structural integrity. Constr Build Mater 167. Elsevier: 846–852

Davis J, Edgar T, Porter J, Bernaden J, Sarli M (2012) Smart manufacturing, manufacturing intelligence and demand-dynamic performance. Comput & Chem Eng 47. Elsevier: 145–156
de Sousa Jabbour ABL, Jabbour CJC, Godinho Filho M, Roubaud D (2018) Industry 4.0 and the circular economy: a proposed research agenda and original roadmap for sustainable operations. Glob J Bus Res 270(1). Springer: 273–286
de Sousa Jabbour ABL, Ndubisi NO, Seles BMRP (2020) Sustainable development in asian manufacturing smes: progress and directions. Int J Prod Econ 225. Elsevier: 107567
Dengiz B, Akbay KS (2000) Computer simulation of a PCB production line: metamodeling approach. Int J Prod Econ 63(2). Elsevier: 195–205
Despeisse M, Yang M, Evans S, Ford S, Minshall T (2017) Sustainable value roadmapping framework for additive manufacturing. Procedia CIRP 61. Elsevier: 594–599
Dillon AP, Shingo S (1985) A revolution in manufacturing: the SMED System. CRC Press
Esmaeilian B, Wang B, Lewis K, Duarte F, Ratti C, Behdad S (2018) The future of waste management in smart and sustainable cities: a review and concept paper. Waste Manag 81. Elsevier: 177–195
Fatemi S, Franchetti MJ (2016) An application of sustainable lean and green strategy with a six sigma approach on a manufacturing system. Int J Six Sigma Compet Advant 10(1). Inderscience Publishers (IEL): 62–75
Garza-Reyes JA (2015) Lean and green–a systematic review of the state of the art literature. J Clean Prod 102. Elsevier: 18–29
Geissdoerfer M, Savaget P, Bocken NMP, Jan Hultink E (2017) The circular economy–a new sustainability paradigm? J Clean Prod 143. Elsevier: 757–768
Geng Y, Doberstein B (2008) Developing the circular economy in china: challenges and opportunities for achieving 'leapfrog development'. Int J Sustain Dev & World Ecol 15(3). Taylor & Francis: 231–239
Ghobakhloo M, Fathi M (2019) Corporate survival in industry 4.0 era: the enabling role of lean-digitized manufacturing. J. Manuf. Technol. Manag.. Emerald Publishing Limited
Harrison J (2006) Six sigma vs. lean manufacturing: which is right for your company. Foundry Management & Technology 13(7):31–32
Hashim S, Habidin NF, Conding J, Fadzlin Mohd Zubir A, NASL Jaya (2012) The integrated between total production maintenance practices and kaizen event practices in Malaysian automotive industry. Int J Eng Res Appl 2(5). Citeseer: 136–143
Hatzivasilis G, Askoxylakis I, Alexandris G, Anicic D, Bröring A, Kulkarni A, Fysarakis K, Spanoudakis G (2018) The interoperability of things: interoperable solutions as an enabler for iot and web 3.0. In 2018 IEEE 23rd International Workshop on Computer Aided Modeling and Design of Communication Links and Networks (CAMAD), 1–7. IEEE
Ioppolo G, Cucurachi S, Salomone R, Saija G, Ciraolo L (2014) Industrial ecology and environmental lean management: lights and shadows. Sustainability 6(9). Multidisciplinary Digital Publishing Institute: 6362–6376
Ivanov D, Dolgui A, Sokolov B (2019) The impact of digital technology and industry 4.0 on the ripple effect and supply chain risk analytics. Int J Prod Res 57(3). Taylor & Francis: 829–846
Kirchherr J, Reike D, Hekkert M (2017) Conceptualizing the circular economy: an analysis of 114 definitions. *Resources, Conservation & Recycling*, 221–232
Kolberg D, Zühlke D (2015) Lean automation enabled by industry 4.0 technologies. IFAC-PapersOnLine 28(3):1870–1875. https://doi.org/10.1016/j.ifacol.2015.06.359
Korhonen J, Honkasalo A, Seppälä J (2018) Circular economy: the concept and its limitations. Ecol Econ 143:37–46
Kravchenko M, Pigosso DCA, McAloone TC (2020) Circular Economy enabled by additive manufacturing: potential opportunities and key sustainability aspects. DS 101: Proceedings of NordDesign 2020, Lyngby, Denmark, 12th-14th August 2020, 1–14
Kuik SS, Kaihara T, Fujii T (2016) Product recovery configuration decisions for achieving sustainable manufacturing. Procedia CIRP 41. Elsevier: 258–263
Kusiak A (2018) Smart manufacturing. Int J Prod Res 56(1–2). Taylor & Francis: 508–517

Lüdeke F, Gold S, Bocken NM (2019) A review and typology of circular economy business model patterns. J Ind Ecol 23(1):36–61
MacArthur E (2013) Towards the circular economy. J Ind Ecol 2:23–44
Millar N, McLaughlin E, Börger T (2019) The Circular Economy: Swings and Roundabouts? Ecol Econ 158:11–19
Moeuf A, Pellerin R, Lamouri S, Tamayo-Giraldo S, Barbaray R (2018) The industrial management of smes in the era of industry 4.0. Int J Prod Res 56(3). Taylor & Francis: 1118–1136
Murray A, Skene K, Haynes K (2017) The circular economy: an interdisciplinary exploration of the concept and application in a global context. J Bus Ethics 140(3). Springer: 369–380
Nadeem SP (2019) Coalescing the lean and circular economy. IEOM Society
Okorie O, Salonitis K, Charnley F, Moreno M, Turner C, Tiwari A (2018) Digitisation and the circular economy: a review of current research and future trends. Energies 11(11). Multidisciplinary Digital Publishing Institute: 3009
Furukawa PDO, Cunha ICKO, Pedreira, MDLG (2016) Evaluation of environmentally sustainable actions in the medication process. Rev Bras Enferm 69(1). Associação Brasileira de Enfermagem: 16–22
Paris H, Mandil G (2017) Environmental impact assessment of an innovative strategy based on an additive and subtractive manufacturing combination. J Clean Prod 164. Elsevier: 508–523
Pavnaskar SJ, Gershenson JK, Jambekar AB (2003) Classification scheme for lean manufacturing tools. Int J Prod Res 41(13). Taylor & Francis: 3075–3090
Pepper MPJ, Spedding TA (2010) The evolution of lean six sigma. Int J Qual & Reliab Manag. Emerald Group Publishing Limited
Pieroni MP, McAloone TC, Pigosso DC (2019) Business model innovation for circular economy and sustainability: a review of approaches. J Clean Prod 215:198–216
Rajput S, Singh SP (2019) Connecting circular economy and industry 4.0. Int J Inf Manage 49:98–113
Reche AYU, Junior OC, Estorilio CCA, Rudek M (2020) Integrated product development process and green supply chain management: contributions, limitations and applications. J Clean Prod 249. Elsevier: 119429
Romero CAT, Castro DF, Ortiz JH, Khalaf OI, Vargas MA (2021) Synergy between circular economy and industry 4.0: a literature review. Sustainability 13(8). Multidisciplinary Digital Publishing Institute: 4331
Romero D, Rossi M (2017) Towards circular lean product-service systems. Procedia CIRP 64. Elsevier: 13–18
Rönnlund I, Reuter M, Horn S, Aho J, Aho M, Päällysaho M, Ylimäki L, Pursula T (2016) Eco-efficiency indicator framework implemented in the metallurgical industry: part 1—a comprehensive view and benchmark. Int J Life Cycle Assess 21(10). Springer: 1473–1500
Rosa P, Sassanelli C, Urbinati A, Chiaroni D, Terzi S (2020) Assessing relations between circular economy and industry 4.0: a systematic literature review. Int J Prod Res 58(6). Taylor & Francis: 1662–1687. doi:https://doi.org/10.1080/00207543.2019.1680896
Rother M, Shook J (2003) Learning to see: value stream mapping to add value and eliminate muda. Lean Enterprise Institute
Russell RS, Taylor BW (1999) Operations management, 2nd Edn. Prentice Hall, Upper Saddle River, NJ
Rymaszewska A, Helo P, Gunasekaran A (2017) IoT powered servitization of manufacturing–an exploratory case study. Int J Prod Econ 192. Elsevier: 92–105
Sanders A, Elangeswaran C, Wulfsberg J (2016) Industry 4.0 implies lean manufacturing: research activities in industry 4.0 function as enablers for lean manufacturing. J Ind Eng Manag 9(3):811–833. https://doi.org/10.3926/jiem.1940
Schäfers P, Walther A (2017) Modelling circular material flow and the consequences for SCM and PPC. Glob J Bus Res 11(2):91–100

Stahel WR (1997) The service economy: 'wealth without resource consumption'? Philosophical Transactions of the Royal Society of London. Ser A: Math, Phys Eng Sci 355 (1728). The Royal Society: 1309–1319

Stock T, Seliger G (2016) Opportunities of sustainable manufacturing in industry 4.0. *Procedia Cirp* 40. Elsevier: 536–541

Suárez-Eiroa B, Fernández E, Méndez-Martínez G, Soto-Oñate D (2019) Operational principles of circular economy for sustainable development: linking theory and practice. J Clean Prod 214. Elsevier: 952–961

Sundar R, Balaji AN, Satheesh Kumar RM (2014) A Review on Lean Manufacturing Implementation Techniques. Procedia Eng 97. Elsevier: 1875–1885

Taddeo R, Simboli A, Di Vincenzo F, Ioppolo G (2019) A Bibliometric and Network Analysis of Lean and Clean (Er) Production Research (1990/2017). Sci Total Environ 653. Elsevier: 765–775

Tao F, Qi Q, Liu A, Kusiak A (2018) Data-driven smart manufacturing. J Manuf Syst 48 (July). Soc Manuf Eng: 157–169. doi:https://doi.org/10.1016/j.jmsy.2018.01.006

TEULON H, (2015) Le Guide de l'Eco-Innovation. Eyrolles, Paris

Wang J, Ma Y, Zhang L, Gao RX, Wu D (2018) deep learning for smart manufacturing: methods and applications. J Manuf Syst 48. Elsevier: 144–156

Conclusion

The ambition of this book is to enlighten SMEs on the approaches available in the field of sustainable excellence. Using both an empirical and practical approach, this book constitutes a relevant reference for SMEs managers seeking to manage their operations sustainably, efficiently and resiliently.

We also integrated the smart component to the sustainable business excellence and proved for those who are still skeptical that SMEs could benefit from smart and digital technology in the favor of sustainability and business excellence. Overall, we build on successful initiatives we experienced in real-life to guide SMEs in order to be sustainable smart and resilient in the post COVID context which is highly demanding in terms of business excellence and efficiency.

Indeed, sustainable excellence provides a competitive position and advantage when decision-makers choose particularly relevant approaches. Approaches allowing the integration of stakeholders in the corporate strategy and business excellence programs such as quality management, lean manufacturing, the six sigma and Industry 4.0 that are frequently studied in terms of sustainable performance.

Also, the proposed improvement approaches are not developed in isolation with the knowledge and practices known by the company, it is simply a strengthening and complementarity view with new horizons of development. Especially in a rapidly changing environment where the survival of an organization depends not only on its most cost-effective, effective and efficient operation, but also its compromise with environmental regulations and social requirements and the adoption of evolutionary strategies. As a result, sustainability of operations appears to be a relevant challenge for organizations.

It is an opportunity to rethink operational practices under a vision of general interest. Considering the three dimensions of sustainable development will propel companies into a sphere of innovation and creation.

F. E. Touriki et al., *Sustainable Excellence in Small and Medium Sized Enterprises*, Industrial Ecology, https://doi.org/10.1007/978-981-19-0371-7

Lightning Source UK Ltd.
Milton Keynes UK
UKHW020653210322
400371UK00001B/3

9 789811 903700